MEMORIES OF A MACKINAC NATIVE

LIFE ON THE ISLAND FROM 1940s TO 2020s

Tom Chambers

2nd Printing – October 2024

Modern History Press

Ann Arbor, MI

Memories of a Mackinac Native: Life on the Island from 1940s to 2020s

Copyright © 2024 by Tom Chambers. All Rights Reserved

2nd Printing – October 2024

ISBN 978-1-61599-831-9 paperback
ISBN 978-1-61599-832-6 hardcover
ISBN 978-1-61599-833-3 eBook

Published by
Modern History Press   info@ModernHistoryPress.com
5145 Pontiac Trail   tollfree 888-761-6268
Ann Arbor, MI 48105   fax 734-663-6861

Distributed by Ingram Content Group (USA, UK, EU, AU)

Cover: Tom Chambers photo (2021) overlooking the town from Fort Mackinac, with the freighter *Stewart J. Cort* going through.

Library of Congress Cataloging-in-Publication Data

Names: Chambers, Tom, 1953- author.
Title: Memories of a Mackinac native : life on the Island from 1940s to 2020s / Tom Chambers.
Other titles: Life on the Island from 1940s to 2020s
Description: Ann Arbor, MI : Modern History Press, [2024] | Summary: "Tom Chambers details his life growing up and career on Mackinac Island, Michigan including the growth of tourism through the building of the Mackinac Bridge, the increasing size and speed of passenger ferries, new transportation modes on the island, as well as making a living through a patchwork of trades including music"-- Provided by publisher.
Identifiers: LCCN 2024031984 (print) | LCCN 2024031985 (ebook) | ISBN 9781615998319 (paperback) | ISBN 9781615998326 (hardcover) | ISBN 9781615998333 (ebook)
Subjects: LCSH: Chambers, Tom, 1953- | Mackinac Island (Mich. : Island)--Biography. | Mackinac Island (Mich. : Island)--Social life and customs. | Mackinac Island (Mich. : Island)--History. | Island life--Michigan--Mackinac Island (Island)
Classification: LCC F572.M16 C53 2024 (print) | LCC F572.M16 (ebook) | DDC 977.4092 [B]--dc23/eng/20240820
LC record available at https://lccn.loc.gov/2024031984
LC ebook record available at https://lccn.loc.gov/2024031985

Contents

List of Photos iii

Acknowledgments 1

Mackinac of the 1940s and 1950s 2

Family Background 8

Early Years 17

Elementary School Years 20

High School Years 23

Following the Ferries 28

Traveling By Air 40

When 10-Speeds Came to Mackinac Island 42

When Snowmobiles Came To Mackinac Island 45

First Band, 1969-73 49

The (Old) Village Inn: 1969-72 61

Silver Birches 65

Summers of 1973-74 70

Bike Races 72

Working For "The General" 76

Mackinac Living, 1973-1980 80

Dabbling in Firearms 84

The Day the Boats Almost Quit 85

Travel 1978-79 88

Trying My Hand at Cooking 91

Iroquois on the Beach: Working For the Macs, 1983-1993 97

Mississippi River Trips 100

A Lifelong Love of Music 107

Golf, 1967-2001	110
Playing Games	112
A Return To Music, 2008-2022	116
Is Mackinac Haunted?	120
Mackinac Island Trivia	122
Model Ships	131
Close Calls	133
Ste. Anne's Church Square Dances	135
Postscript	137
About the Author	138

List of Photos

City of Detroit III at Mackinac Island (1949) Photo by Tom Chambers' mom 3
Main Street slide winter 1959-60. Photo by Jack Chambers 5
Great Uncle Captain Tom Chambers 9
Mom on her cycle truck (1952) 9
Great Grandfather Alexander Duffina (1835-1926) 10
Jack's Victoria Buggy (1954) 10
Jack Chambers and Uncle Dennis Dufina with the Lilac Queen, mid 1950s 11
Our Victoria carriage 1954. Photo by LL Cook 11
Jack Chambers with two of his horses on Bogan Lane (1955) 14
Dad and Grandparents dinner at Iroquois Hotel (1960) Photo by Frye 14
Somewhere in Time party Mom Ray and sister Kathy with the stars (1979). Photo by Sandi Brady 16
Dock near our house 1949, built by jack & Matt Yoder. photo by my mom 16
4th grade on the Island 22
Straits of Mackinac II at Straits Transit dock (1970) 32
Shepler's *Wyandot* (2017). Photo by Tom Chambers 36
Mackinac Island Airport (1966) 41
Snowmobiling on the ice (1988) 46
Morningstar band, beach party (September 1973) 58
Ray Miller and Jim Enrietti of the house band The Zippy Zips 62
David, Matt and Tom revisit our 1971 home about 15 years later 66
1975 relay race. Ty's Restaurant team 73
Painting of my grandfather Phil's cabin at British Landing. The dog's name is Patches 81
Cooking at Holiday Inn Gulfview 1980 93
Star Line Cadillac in St Louis (May 1990) 105
Playing a board game (1985) 113
Chimney Rock band (08/20/2017) 118
The abandoned 'Haunted' Harvey House. Photo by Loren Horn 121
Four of Tom Chambers' more than 600 hand-built model freighters 132
A typical square dance, with jimmy Boynton calling (undated) 135
Mackinac Island Trail Map by Chris Bessert 139

iii

Acknowledgments

Maybe I have always wanted to write. I got my start trying my hand at short stories as a teenager, and I imagined writing a book. I took one or two writing classes each of the five years I went to college. They varied from a 500-level creative writing class to something a bit esoteric called "Structure of Modern English" where you learned advanced diagramming and proper usage. Examples would be, which is correct, "A" history or "AN history"? Or should I say, "that weighs a ton!" or "That weighs such a ton!" Suffice it to say, my college classes didn't produce any bestsellers.

As for writing memoirs about a place you know well, a lot can be said for observing, paying attention, reading, remembering, and being able to string a grammatically correct sentence together. I wrestled with how much should be history, and how much autobiography.

I acknowledge Mackinac writers who came before me: my mother, Mary G. Duffina, Tom Pfeiffelmann, Jack Welcher, Dr. Eugene Petersen, Phil Porter, Dennis Cawthorne, Marta Olson, Dr. David Armour, and Edwin O. Wood.

A nice thing about this volume, if a chapter isn't your cup of tea, or it's too much "inside baseball," you can simply skip it, and move on to the next. Each chapter is labeled.

I frequently found myself waking up in the middle of the night with ideas these past few months, and I used up most of a pack of printer paper during trial and error. I have tried to mention all the memorable people who have passed through my life. I apologize if I have forgotten anyone, but of course some people are not included by name, for privacy reasons. My thanks to all of you, for your interest in reading it.

In 1972, one of my English teachers at North Central Michigan College, the late Tusco Heath, wrote in my sophomore yearbook, "Finish that book, or I'll haunt you!" Well, it may have taken me fifty years, Dr. Heath, but here it is.

2nd Printing, October 2024
Mackinac Island, Michigan

Mackinac of the 1940s and 1950s

For this memoir, I will not attempt to cover a detailed Mackinac Island history. Others have already done that, and better than I could. I will give just a bit of background on the decade I was born, and the one preceding it. Visitors always like to comment that Mackinac Island "looks the same" as it did in a previous time, but I notice changes every few years or even every year. Every few decades there is an earthshaking change that really alters the complexion of the Island.

One of the biggest differences was the means of getting to the Island. From about 1870 through the early 1950s, the primary way was via large steamship, originating at ports such as Detroit, Cleveland and Chicago. There were several major steamship lines, and over twenty vessels, which stopped at the Island, docking at the Arnold Dock and Coal Dock. Ships also had routes to Sault Ste Marie and Lake Superior, after the Soo Locks were built. The lines were Northern Steamship Co., Anchor Line, Lake Michigan & Lake Superior Transportation Co., Northern Michigan Transportation Co., Cleveland & Buffalo Transit Co., Great Lakes Transit, Detroit & Cleveland Navigation Co., (they had the most ships), Georgian Bay Transit, and a rare trip by Canada Steamship Lines.

My grandfather, Ed, visited during most summers in the 1950s, taking the SS North American from Chicago. The highway system was primitive in those days, as were cars. Before the Mackinac Bridge was built in the mid-1950s, the only way across the Straits of Mackinac way by large white State of Michigan car-ferries. Airlines didn't really serve our area, but there were two Air Force bases in the Upper Peninsula. As highways got better, the era of the steamers went into a slow decline, in the late 1940s. Detroit & Cleveland Line's massive "Greater Detroit" last called here in 1950.

From 1951 on, it was just the *North* and *South Americans*. The *North's* last season was 1964, and the *South's*, 1967. The big game-changer came in the late 1950s, when President Ike

City of Detroit III at Mackinac Island (1949) Photo by Tom Chambers' mom

Eisenhower introduced the Interstate Highway system of modern 4-lane divided highways. Travel and tourism in our area benefited greatly by the north/south running "I-75."

There were small high-speed ferries running to the Island then, but they were of natural mahogany instead of aluminum. Most of them docked at small wooden docks between the Coal Dock and Iroquois. These all slowly went out of business, the last one being the 48' *Fairy Isle*, which moored at the Arnold Dock until the mid 1960s. Their legacy was carried on by William "Cap" Shepler with his 30' and 36' wooden speedboats, Miss Margy and Billy Dick. They docked at the Welch/Straits Transit dock on the east end of town.

Home heating was also much different in the 1940s and 1950s. Nobody had electric heat. There were oil stoves and oil furnaces, coal stoves and coal furnaces, fireplaces and wood stoves. It was always nice walking around town on a winter evening and smelling wood smoke coming out of the chimneys. If you had a central oil or coal furnace, usually in the basement, chances are the rooms of the house each contained a radiator, filled with hot water and having a valve at the bottom and a steam release nozzle at the top. I recall my great aunt Ann and uncle Tom had large Jungers oil stoves in two rooms. These were about the size of a refrigerator and had a small glass door where you could see a glowing orange burner. Drays would deliver oil for furnaces carried in a four-wheeled tank wagon. Oil was then pumped by hose to the furnace. It was similar to gassing up your car.

If you burned coal, drays also delivered that, from the Coal Dock. There were two heaping piles of coal on the dock, one hard coal (anthracite) and one soft coal (bituminous). Soft coal chunks were about the size of a softball and burned in open stoves. Hard coal was just smaller than a golf ball, and burned in a furnace. The coal was shoveled down a coal chute into the basement (these could be seen all over town). Somebody in the household then had the job of "stoking" the coal furnace, shoveling coal into it, as you would on a steamer type ship. Later, waste product, "clinkers" had to be removed from the furnace when it was cool. I remember having clinker duty at our big two-story house at British Landing, 1966-67.

Wood for your fireplace or kitchen wood stove could be delivered by drayload also. It was often pre-split and delivered in quantities called a cord, which was equal to 128 cubic feet cordwood. It was still possible to go out and cut your own wood, as in the very old days, either on Mackinac or Bois Blanc Island.

It goes without saying there were no air conditioners in town those days either. Even in the early 1970s they were rare. Now, in the 2020s, every summer employee has one.

There were a lot more trees on Market Street and in town. Green space was more abundant too, even into the 1960s. In addition to the small park behind the Carriage Tours booth, there was one at the head of the Arnold Dock, between the Chuck Wagon and McNally's, and west of the liquor store, all the way to Iroquois.

The types of shops along Main Street have changed over time. Around the turn of the century, there were a half dozen rug merchant establishments. In the early 20th Century, you could find a dentist's office and three drug stores: Bogan's, Bailey's, and Central Drug Store. Some of them, especially Central, included a "fountain" counter, with seats for customers to get beverages. The type of gift shops has changed over time also. I feel in general there were more quality shops when I was growing up, than there are now.

I often patronized Bill Cooper's Balsam shop to buy fancy toy soldiers, and his wife Doris' Copper's Perfume Shop, which was much more than perfume. She also carried a variety of turquoise jewelry. Maria Moeller's Imports shop was also excellent. When the Doud family bought it and renamed it Windermere Imports, it carried on the tradition. There were a few shops in town that carried the famous Hummel figurines. I remember Frank Shama's did. Clarice McKeever

Main Street slide winter 1959-60. Photo by Jack Chambers

Haynes had a nice art studio. At least there is still an art gallery today, the Little Gallery on Market Street.

In the early 1900s, merchants from all over the world found their way to Mackinac. In the 1950s and 60s there were a few ethnic restaurants along Main Street. Harry Stamas had a Greek Restaurant, and Sam Brocato had an Italian one, with a dancing marionette in the window. My favorite was Iggy and Catherine Palermo's "Chatterbox" restaurant, located where the Lilac Tree courtyard is now. It featured homemade Italian bread and Spaghetti dinners. My family also patronized Wandrie's Restaurant in late 1959 and early 1960s. My godfather and family friend, Dennis Brodeur, was running it. The chicken dinners were spectacular. Little Bob's on Astor Street was also popular, with Islanders and tourists alike. It was known for the giant cinnamon roll.

Some other restaurants I remember, the Buggy Whip and the Carriage Lantern, went out of business. I also recall the Arab and Jewish merchants along Main Street, Edward and May Sultan of Edward's Gifts, and Roben Arbib, who had The Thunderbird at the head of the Arnold Dock. Tom Shamy had a linen shop for many years in the current Mighty Mac location. There were far less fudge shops in the old days, too, and two shops from the 1960s even were bought out and renamed, these being Mary's Fudge owned by Florian Czarnecki and Suzanne's Fudge owned by Mary and Ray Summerfield. Frank Nephew acquired them both, and they are now Joann's Fudge. Nowadays, Main Street is dominated by Sweatshirt and Tee-shirt shops, and of course fudge. Somewhere along the line the term "Junk Shop" came into being. It referred to the newer, low-end gift and souvenir shops.

Some quality clothing stores in the 1950s went out of business. "The Scotch House" in the Murray Hotel building featured wool and plaid Pendleton clothing. It was run by Joe and Letty Abbey. "Neary's" was located where the Pancake House is now, and it had quality apparel. The family had a house on the East Bluff, and would winter in Arizona. Before Neary's the whole two-shop Dennany building was Herpolsheimer's Department Store, in the 1940s. I also remember "Angel's" clothing store in the Chippewa Hotel building. It was run by mustachioed Freddie Ashton, who always sported an ascot.

Boat docks along the waterfront were everchanging as well. In the old days, the big Arnold/Star Line dock was called Hoban's wharf. The Coal Dock was (James) Bennett's wharf, and for many years it had a large steam crane on it. There was also a third large dock, just east of the Arnold Dock, called the Government Dock. It was quite active in the 1870s, but late in the 1800s it was torn down. Some piling stumps remained for several years.

In the 1930s and 40s, there were several small wooden docks between the Coal Dock and Iroquois Hotel for the high-speed ferries, in the age of the small mahogany cruisers. They included the Couchois Dock, Bird's Dock, and others. They were all discontinued at the end of the 1940s, and that area was empty beach for twenty years. Shepler's started building their "new" dock in 1966, and Harry Ryba built a dock across from the Lake View in 1967-68. There was a medium-sized dock behind Bennett Hall until around 1950. The dock behind Bay View Cottage went through many names in 120 years. The first version was called the Bailey dock, after the family residing at Bay View. Off and on it was known as the Yoder Dock, for another Bay View Resident.

The old dock was completely torn down in the mid 1940s and there was no dock there again until 1949. In most of the 1950s, it was known as the Welch Dock, for Mackinaw City ferry operator Dick Welch. From 1958 until the mid-1980s, it was the Straits (Transit)

Dock. Straits had lengthened it during their tenure, but it was still all wood. Arnold put in a new dock in the mid-1980s, with steel I-beam pilings and a wooden deck. In the 1990s, Arnold Transit began calling it the "East Dock," I suspect to erase the memory of their conquered competitor, Straits Transit. A railing was installed the length of the dock, down the center, as Doug Yoder maintained ownership of the west side. "East Dock" never caught on with me, both because I was a Straits fan, and also because strictly speaking the Beaver Dock was the easternmost dock in town. It had been built by Moral Re-Armament in the 1950s, in the shadow of the East Breakwall. It was used to load and unload freight aboard their 65' powered barge, Beaver, for their myriad projects.

In 1949, two Island ladies, Nurse Stella King and Evangeline "Ling" Horn, came up with the idea of honoring our prolific flower, the lilac tree. A Lilac Parade was born, with horse-drawn floats and marching bands, held on a Sunday afternoon in mid-June, to coincide with the lilacs blooming. Sue Perault was the first Lilac Queen. A Lilac Coronation Ball was also held at the Community Hall the Saturday night before the parade. The Queen held a "court" of four or five ladies, and she was crowned during the Ball by a local or State dignitary who was also emcee of the night's dancing and coronation.

Some later Lilac Queens were Mary Dennany in 1950, Jeanie Vance in 1951, Catherine McGreevy in 1952, and Joann Goodhart in 1953. There was also musical entertainment each year. The Lilac Festival and Parade continues to this day, and has become a sprawling 10-day event. Sadly, the Coronation Ball was discontinued over thirty-five years ago, much to my dismay. There was rumor some bar owners felt the Ball was taking business away from them, for one night of the entire season. The Queen is now crowned at a small, low-key event in Marquette Park or on Windermere Point, usually by the Mayor. The Ball had been something that really honored the girls, and I'm sure many remembered it for the rest of their lives.

Family Background

Both sides of my family have a long history on Mackinac Island. My mother, Mary G. Duffina, was of French/German/Native ancestry. She traced membership to the Mackinac band of Chippewa Indians, and her family name was included in The Durant Roll, a census taken by the U.S. Dept. of the Interior in 1907-1910 to determine tribal membership. She was born on Mackinac Island, August 8, 1929, eldest daughter to Philip G. Dufina from the Island, and Elizabeth Habermehl Dufina, formerly of Alpena. The time of her youth was divided between Mackinac Island and Alpena. She had three younger siblings, sisters Lois and Frances, and a brother, Dennis, the youngest. She played clarinet at Alpena High School.

Her father, Phil (1899-1954) was a scratch golfer, and was golf professional at the Grand Hotel for a few years. He had a variety of trick golf shots he demonstrated to local golf fans. All six Dufina brothers of that generation were avid golfers. Phil's brother, Jim succeeded him as pro at The Grand. My mom's mother, Beth, was a cook on the Island for many years. My mom developed an interest in horses at an early age, and she was involved in my dad's carriage and buggy services for about ten years. She was married to him for fourteen years. From the mid-1950s to mid-1960s, she was in the Ste Anne's Church choir, along with Mary Kate McGreevy, Gloria Davenport, and Nova Therrien. She also learned to play the electric organ, as Mary Kate did. My mother always put a lot of trust in authority figures—doctors, law enforcement and the clergy.

In 1962, she married her second husband, Albert R. "Ray" Summerfield, who came here as foreman for the construction of the new Public School. They had four kids between 1963 and 1968: David, Suzanne, Sheryl, and Kathryn. Mom was also involved in some downtown business ventures in the mid- to late-1960s, with Ray. For a year in the early 1970s, she became interior decorator for the Rev. Rex Humbard's Stonecliffe. Mom and Ray were married for twenty-three years. She also wrote two books about Mackinac Island.

Many of the Dufinas are buried at Ste. Anne's Cemetery on the Island. A few years ago, I located the gravestones of my great-great-

Great Uncle Captain Tom Chambers

Mom on her cycle truck (1952)

Great Grandfather Alexander Duffina (1835-1926)

Jack's Victoria Buggy (1954)

Jack Chambers and Uncle Dennis Dufina with the Lilac Queen, mid 1950s

Our Victoria carriage 1954. Photo by LL Cook

grandfather, Alexander, born in 1835, and his wife, Ursula, born in 1844. There are probably even older ancestors there.

My father, John Thomas "Jack" Chambers, born Jan. 29, 1926, was of Irish descent. His ancestors came to Michigan during the Irish Potato Famine of 1845-52. There were four branches of the Chambers family in the area, three on the Island and one in St. Ignace. Jack was born in Chicago, the second son of Edward and Mary (Cullinan) Chambers. He had a brother, Bill, nicknamed "Will'am," two years older. Mary died of an illness when Jack was about six years old.

Jack's grandfather, "Cannonball" Bill Chambers, had settled on Mackinac Island and bought a few acres of land at British Landing. He was so-named because while digging on his property, he unearthed a 33-pound historic cannonball, probably from the War of 1812. He raised a family there at his store and residence, called aptly enough, The Cannonball. He also had a large vegetable garden, and sold his produce in town. Cannonball Bill had three kids, Edward (b.1892), Thomas (b.1894) and Ann (b.1896). Edward eventually moved to Chicago, and in the early 1930s, Tom and Ann moved to town, running a rooming

house they called "The Ivanhoe." It is now owned by the Gough family, and known as "The Big House."

After their mother died, Jack and Uncle Bill were sent to the Island to live with Tom and Ann. From the 1930s through the 1950s, Tom had a seasonal career sailing on Great Lakes freighters, and later yachts. He eventually became a captain. He even spent some time on the train ferry, *Chief Wawatam*. Around 1939-40, Tom and Ann purchased a house from the Dr. Bailey family, on the east end of town, called "Thuya." Ann ran a tourist home (B&B) there in summers. In the off-season she was school cook at Thomas W. Ferry School. Everyone raved about her meals and desserts. Tom and Ann also inherited the cannonball from their father when he passed. I'm sure the State Park would have loved to acquire it. When Tom and Ann moved to St. Ignace in 1969, they gave me the cannonball.

Jack Chambers entered the merchant marine in 1944 at eighteen, and made trips to Europe and the North African port Mers-El-Kebir. He served on the Liberty ship Marine Panther, among others. At the end of the war, he made a couple of trips on a liberty ship to South America. In 1946, he returned to the States and did a short term as junior engineer on the Great Lakes tanker, Beaumont Parks. By summer

1946, he was back on Mackinac Island, becoming a dockporter for Chippewa Hotel's new owner, Nathan Shayne. He and my mom married at Ste. Anne's Church in summer of 1946. Jack's brother, Bill, was best man, and Sally Chambers was maid of honor.

My parents soon developed an interest in horses and got into the milk delivery business. Jack used a single horse hitch wagon. Sometimes his brother William helped out. My mom got a Schwinn "Cycle Truck" bike with an oversized basket, and would make all sorts of deliveries with it.

They bought a lot at the top of Bogan Lane, west side to pasture horses. Soon, Jack built a fairly large two-story barn there as well. Jack took a brief hiatus from the horse business to work for Moral Re-Armament (MRA) in the early 1950s as many Islanders did. He was chiefly involved in the building of the theater, as a foreman along with Chuck Dufina and Jim Francis. Jack was particularly proud of his accomplishment building the theater roof and ceiling, with its large log beams.

After about a year he became disenchanted with MRA, and got back into horses, starting a livery tour business. He had two primary fancy carriages, a black open top Victoria buggy with two facing seats and a "dinky" seat for the driver, and a varnished natural birch wood carriage he and Uncle Dennis Dufina built. It also had facing seats, and later had a roof added. His carriage service employed three drivers: himself, Dennis and a fellow named Don Courtright. Jack outfitted them in smart uniforms, consisting of a long yellow coat, bright green pants, top hat and puttee-style black boots. The overall appearance of the rig and driver was similar to Grand Hotel's vis-a-vis carriage, and was, I felt, the best-looking in town. My mom did the bookwork for the business. Pro photographer, L.L. Cook, took several postcard photos of Jack's fancy rigs and drivers in front of the Park, at Grand Hotel, and behind the Fort.

The livery venture lasted three or four summers before Jack got into management of the newly formed Carriage Tours Corporation, which had consolidated all the independent livery drivers in town, as well as having large three and four seat buggies to carry tourists. Jack became vice president of Carriage Tours under his cousin, Arthur T. Chambers Sr., and held the position about five years. He organized the Carriage Tours reservations system for loading tour buggies on Main Street. He remained a consultant for Carriage Tours for the rest of his life.

Jack Chambers with two of his horses on Bogan Lane (1955)

Dad and Grandparents dinner at Iroquois Hotel (1960) Photo by Frye

Jack got into the audio-visual business for about a year, in 1962-63. He had always been an avid slide photographer. My mom and dad divorced in 1959, and Jack spent winters in Mexico for a few years, first Mexico City, and then Acapulco. He became interested in Mexico through Island businessman Tony Trayser. During the 1960s, many Mackinac business people developed an affinity for spending winters, or winter vacations there, including Otto and Marge Lang, Louie Deroshia, and Gene and Marian Petersen. In 1964, Jack went into partnership with Dennis Brodeur of Wandrie's Restaurant. Dennis was also my godfather. They bought a couple of Island businesses from Bob and Arlene Chambers (no relation) of Florida.

The businesses were The Orpheum Theater, which they owned for about six years, and The Village Inn, a narrow two-story bar and restaurant located between McNally's and the Lake View Hotel. My first job was at the Orpheum, and Jack's second wife, Joyce, sometimes worked in the ticket window, rotating shifts with Marilyn or Carolyn Riley. I discuss The Village Inn with more detail in another chapter.

After my dad and mom divorced, Jack moved into a small house on Bogan Lane, next to his barn, which I believe had belonged to his cousin, Lucy Mary Chambers. He built an addition on the back and modernized the kitchen, with hardwood floors and varnished wooden cabinets he built himself. I remember he rode a blue Schwinn Tiger 3-speed middleweight bike with a large basket. Once on a rainy day, his hand brakes failed stop him coming down Bogan Lane, and he crashed across the road, vowing never to ride a bike with handbrakes again. After that he switched to middleweight Schwinns with the two-speed kickback hub. A side note, Jack's brother, Bill, was a department head (franchising manager) in Schwinn Bicycles, of Chicago.

Jack Chambers and my Uncle Dennis Dufina were actually brothers-in-law twice, which is quite unusual. The first time was when my mom and dad were married, and Dennis was my mom's brother. The second time, Jack's second marriage was to Joyce Mikesel, younger sister of Dennis' wife, Carol.

Jack Chambers also became building inspector for a time, sat on the City's Board of Review, and worked with Dan Musser in organizing the Department of Public works.

Both my parents left Mackinac for the mainland eventually. My mom, Mary, retired to Florida, where she worked on her writing, and enjoyed riding her yellow Schwinn one-speed bike she had brought from the Island. Jack bought a house in St. Ignace. He passed away on April 1st, 2005, and Mary on September 26th, 2018.

**Somewhere in Time party Mom Ray and sister Kathy with the stars (1979).
Photo by Sandi Brady**

Dock near our house 1949, built by jack & Matt Yoder. photo by my mom

Early Years

I was born on a July night in 1953, in the small one-story house my dad had built next to Sara Yoder's Bay View Cottage. It had gray slate siding and a red roof. Down at the waterfront was the small wood crib dock my dad and Matt Yoder had built a few years before, with the help of Cowell's tractor. The dock was leased to Dick Welch of Mackinaw City, who hauled mail and passengers on his 55-foot vessel, *Buddy-L*. We lived there for four years, including a winter or two. Jack worked for MRA the winter they built the theater, and he was crew foreman for that project. He engineered the hanging of the log beams on the roof. Assistant foremen on the crew were Jim Francis and Chuck Dufina. When we moved, our house was rented to the J. Wellington family. They had a daughter my age, Rosemary.

In 1958, we got an opportunity to rent the two-story house with the red roof across from Ste Anne's Church. My mom knew the owner, a Catholic priest, Fr. Jodesy. I thoroughly enjoyed the five years I lived there. The living room, dining room and den had varnished bead board and wainscoting. There were four bedrooms on second floor, and a full attic with a tin floor in the turret room. The large basement accommodated my O-27-gauge train set on a big sheet of plywood.

In the summers, Jack ran his hourly carriage business, and built a new barn across from Carl Couchois' at the top of Bogan Lane. His drivers were Don Cortwright and my Uncle Dennis. During the winters of 1957 and 1958, we moved to Florida, first to Daytona Beach and next to Sarasota. My dad found work in Sarasota cutting trees and clearing land for the burgeoning growth in the state. I enjoyed stopping at Cypress Gardens on the way down. Uncle Dennis Dufina was in Florida during those winters, too, working as a hotel bellhop. Later he worked at the Belleview Biltmore Golf Course. He developed a fondness for wintering in Florida that lasted a lifetime. While in Daytona, I got to go fishing a few times with Jimmy Chambers.

One of my early Island memories, while browsing in The Big Store, owner Tony Trayser showed me a large diorama model of the Island he had in the back room. It was about eight by ten feet and sat on a tabletop. The 1.5 inch trees all looked realistic, and the model was lighted, mainly in the downtown area.

Also while in Daytona, Florida in the late 1950s, my family spent time with Dennis and Leetha Brodeur and kids Fred and Leanne, who was my age. We used to return to the Island each summer.

My mom and dad divorced in 1959. Mom and I spent the following winter in East Lansing. The autumn after that, we traveled out west, stopping at the Petrified Forest and Tucson's Superstition Mountains, before visiting Aunt Francis in San Diego. Uncle Dennis was out there too, now in the US Army, and stationed at Fort Ord, California.

We decided to spend the rest of the winter in Scottsdale, Arizona, and I was enrolled in third grade there. Scottsdale was real nice back then, and small, about the size of Petoskey. We got an apartment in a small one-story complex that had a swimming pool. Uncle Dennis often visited when on leave, and my dad dropped in on his way to Acapulco, Mexico for the winter. I remember all of us taking a dip in the pool. My dad wanted to take me to Acapulco for a couple of weeks, but my mom vetoed it. I was quite disappointed.

Once again when springtime came, each of us returned to Mackinac Island, where we lived one more year at the Jodesy house. I remember my friends from the Mission and I building rafts when we were 10-11 years old. Most were just logs with planks or plywood on top. Terry Francis had the best one. It was framework, top sides and bottom, and the inside was filled with empty Clorox bottles for buoyancy. In late summer of 1965, about six or eight of us kids gathered up some boxes of apples and went to Ft. Mackinac, divided up into teams, and had an apple fight. It seemed like the perfect location, and we had a lot of fun. Unfortunately, Park Superintendent Carl Nordberg caught us at the end, and made us clean up all the apples.

Mom was married to Ray Summerfield by then, having met him during the new school construction. We would move out in 1963, and the Bloomfield family would occupy it after us. A few years later, Bob Benser rented it. I remember Bob hosting the Nephew-Benser employee parties in the big backyard, where my friends and I used to play. Alice Motillo and her husband, Joe, were the owners now, after Fr. Jodesy passed away. Alice was the late pastor's sister. We rented Bay View cottage from Sarah Bailey Yoder for three years, 1963-65. Sarah used to run a tourist home (B&B) there, but was now elderly, and retired to Quincy, Indiana. Bay View was great for me, since it was at the head of the Straits Transit dock, and I could watch all the boat traffic close up and visit with the crews. Ray built a small party store in the back yard, facing the new yacht dock, and it did great business.

Elementary School Years

When it came time to start school, we were living on the Island. As I recall, for some reason kindergarten was not held in the Thomas W. Ferry School, located between Anne's Cottage and Marquette Park. I remember attending kindergarten in Ste. Anne's Church old basement meeting hall, and also in the upstairs loft, street side in the old Community Hall.

My mom and dad were going through a divorce about this time. So, for 1st grade, my mom moved the two of us to East Lansing in September. She took a job within sight of the capitol, at the tallest building in downtown Lansing, as a secretary. Her skills at shorthand were invaluable in this work. We lived in East Lansing at a small apartment complex on Gunson. St. My school was a three-block walk, at a two-story brick structure called Bailey Elementary School. It looked like your typical 1950s, 1960s elementary school. My teacher was a middle-aged woman with dark hair and glasses, Mrs. Wright. I liked her. I usually tended to get along with teachers throughout my entire education. I still remember some of my first-grade friends at Bailey: Dennis Crittenden, Dale Klinseck, Joanie Summers. My best friend was a girl of Polynesian decent, with short dark hair and bangs, named Leilani Tetsworth. She was the smartest girl in the class, and we were constant companions, be it for homework, or play at one or the other's place. We would also go shopping together at the Frandor Shopping Center in East Lansing.

First grade was one of my favorite school years. Near the end of the year, faculty gave me some evaluation tests, and decided I didn't need to attend 2nd Grade. The promoted me directly to 3rd Grade for the following school year.

About the end of May, Mom and I returned to Mackinac for the summer. We were still renting the Jodesy house, next to Ste Anne's Rectory. I spent the summer, as I had the last one, playing in our big yard and on the beach with my good friends from the Mission, Joey Davenport, Tim Horn, and Terry Francis.

If it was raining, we'd play inside. My dad had bought me a real nice "O" gauge train set, which was set up on a sheet of plywood in the basement. The locomotive made smoke and had a light. There was an "O" gauge town, complete with different colored sandpaper stuff representing grass, sand and other terrain. The whole gang loved the train set. I also had collected different types of toy soldiers, amassing about 600 at my peak. Tim Horn had them also, and his collection numbered 1000. I used to save my money and buy highly detailed medieval knights from Doris Cooper's Perfume Shop or her husband Bill's Balsam Shop. I had about a dozen of these, some on horseback.

As covered in another chapter, I attended 3rd Grade out west in Scottsdale, Arizona. By the time I got to 4th Grade, a new school had been built on the Island, in the shadow of the Grand Hotel, on the Borough Lot. It replaced the Thomas W. Ferry School. It was quite a project, and was planned to accommodate grade K through 10. The 11th and 12th Grades still needed to commute to St. Ignace LaSalle H.S. for a few more years. My mom's new beau and soon to be husband, Ray Summerfield, had been overall foreman on the school building project, under contractor brothers Bruce and Jerry Wadel of Ludington.

The Wadels bought a white LCM-6 landing barge, similar to the one the State Park has. They used it to deliver most of the building materials to the beach below the boardwalk. The barge was named "Gisela." The construction crew numbered around 30 men, and my mom and Ling Horn got the contracts for feeding them. Some ate at the big dining room table in our residence the Jodesy house, and Ling fed hers at Lafayette cottage, across from Mission Church. Ray's brother, Ed, was one of the workers. The Wadels built two new one-story homes on the side street behind Mission Church for their use. One was later sold to Aleda Schmidt.

I returned to school at Mackinac for 4th Grade, in Mary Ravenscroft's room on the southeast corner of the new school. Two grades were combined in each room, so ours was the 3rd and 4th. Plastic still covered the window openings, as the new ones hadn't yet arrived. In those days, there was a high turnover of school superintendents, but the current one, Ken Roberts, had lasted a few years. The following year saw me move to the room across the hall, on the SW corner, for 5th Grade. Our teacher was Fred Joneson, a tall guy with red hair and glasses. He was good friends with the Alford family. By 6th Grade there was a new superintendent, Richard Hendra. His wife taught us in the 5th and 6th Grade room. The following year I moved to the other end of the school for Junior High.

When I was a kid, one of the things I collected were stamps. My collection was modest at first, but it did contain a "First Day of Issue" of the 3cent aqua, Mackinac Bridge stamp. Around 1964, I got a windfall when an English gentleman from MRA named Jervas Reesor decided to gift me his extensive stamp collection. I'm not sure how he even chose me, but he was a friend of family friends who also worked at MRA. My stepdad Ray was doing some work around the Beaver Dock. It was just serendipity, and I was lucky to receive them. Unfortunately, they would burn up in the fire at our British landing house in September 1967.

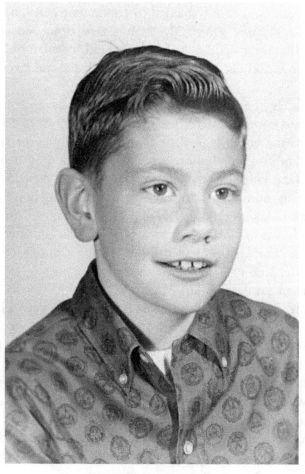

4th grade on the Island

High School Years

For the school year 1964-65, I moved to "The Big End" of the school, as the kids liked to call it. This was for Grades 7-10. Grades 11 and 12 were still commuting to LaSalle High in St. Ignace. My elementary schoolmates Melly Alford and Marie and Gayle Welcher had departed for off-island schools. Melly transferred to Leelanau School near Traverse City, and the Welcher family had moved to Rudyard. The Hendras had departed.

Byron Antcliff was the new superintendent hired by the school board during my 7th grade year. He was tall and athletic, with a dark crew cut and big black glasses. He taught science class, and his wife Kay was also a teacher. They were respected. Longtime teacher and principal, Charles Wellington, retired mid-year, and was replaced by new teacher, Bruce Fillmore. Byron Antcliff had heard the Island boys were a tough lot, especially the village guys, and he wanted to impress upon them the first week that he was no one to fool with. He asked for six volunteers among the guys who thought they were "tough" and they went outside of the high school and had a wrestling match. I preferred to be an observer. I remember guys flying out of the pile left and right. Antcliff held his own, and even won. After that, all the guys respected him for the rest of the year.

He also instituted a running program for physical education class. Everyone ran one lap completely around the Borough Lot. We started up the hill to Mary's Pantry, made the turn paralleling Cadotte Ave, past the Grand tennis courts and the sing ring, and down the hill back to school. Mr. Antcliff ran with us, and even Dr. Joe Solomon participated. I recall Joe Horn being the fastest student, but Mr. Antcliff was no slouch either. The Antcliffs only stayed one year.

During my 8th grade, James Franklin was hired to teach and be superintendent. He came from an agricultural background and had experience with horses. He returned to the Island years later, to drive tour buggy for Carriage Tours. He also had a second tour of duty as superintendent, twenty or so years after he taught me.

In 9th grade, my freshman year, our new superintendent and teacher was Richard Bolander. His wife, Jackie, taught typing. I remember we had just been given gigantic new tan Royal typewriters. Marshall Sheets was also brought in to teach. The Bolanders returned to Mackinac in 2004 after retirement, and bought the former Bob Andress house.

Next year came 10th Grade, my favorite year of high school. New superintendent/teacher was Francis Francomb, and another new teacher was Werner Maul.

A constant, through my grades 8-10 in the teaching department was Dorothy Hart. She was a wonderful teacher, respected and beloved by all the students. She stayed on many years after I had graduated. She was willing to help out with anything, and also chaperoned school dances. Mrs. Hart and her husband, Melvin, had bought the Welcher's house on Market Street when they moved to Rudyard.

10th Grade was a good year in lots of ways. We had extra students, whose parents were involved with Mackinac College. I liked all my teachers, joined the basketball team, and had a girlfriend, a local girl a year younger than me. Mackinac College helped out a lot at school, and also let us use their gymnasium basketball court.

On Halloween, they sent some of their staff to create a wonderful spook house for the kids at the school party. Rusty Wailes, a 6' 6" Olympic athlete, played the Frankenstein monster. I recall him picking Mark Chambers up and giving him a good scare. The college also hosted weekend movies at their theater. They were pretty high-class ones. My girlfriend and I went every week, sometimes with our friends. I remember seeing *Exodus, Becket, Flight of the Phoenix, A Man for All Seasons,* and *Khartoum.* Island girls I was friends with in high school included Debbie Rogers, Laurie Cowell and Debbie De La Verne.

The high school had had a basketball team for a few years, but of course it only included guys up through grade 10. I wasn't on the team in 8th or 9th grades, but I remember James Franklin drove the boys to an away game at Onaway. We played different teams then. There was no Northern Lights League. We also visited St. Ignace to play against their junior varsity team.

I was on the team in my sophomore year and played home and away games with Engadine. Their gym had no heat. My team only had eight players. I played center, Jim Bradley and Terry Francis were forwards, Carl McCready, George Bodwin Steven Green and Jimmy Martin were guards. Bobby Durant played a bit when he was there. He was a cousin of the Cowells, who, along with his two sisters had transferred here from Lawton, Oklahoma.

Our excellent shooting guard from the year before, Bobby Green, was now commuting to St. Ignace. Several good athletes didn't make our team due to academic ineligibility. I recall an away game at Detour (Village). They had 25 guys on their squad to our eight. If players got tired, they just sent in a fresh five. They ended up beating us 108-48, but their scoreboard only had a readout of two digits, so when their score turned over the century mark, Jim Bradley nudged me on the court and said, "Look Cat, we're ahead!" We didn't win many games, but all the boys had fun. Sadly, there was no girls' basketball program yet. There would soon be girls' volleyball.

We also had a spring track team that year and took part in one meet in the Eastern U.P. Bobby Durant was great on the hurdles, and won his race. I recall Billy Fallstich throwing the shot-put. Friday night school dances in our small gym were another fun aspect of 10th Grade. I often brought my LPs to play and share. My girlfriend loved dancing, especially the slow dances.

In high school I got my first job, too. Dennis Brodeur and my dad owned the Orpheum Theater, having recently purchased it from Bob and Arlene Chambers. They hired me to work the concession stand during the summers of 1965-67. My dad's second wife, Joyce, managed the place, Marilyn Riley ran the ticket window, and I recall two guys' named Red and Bob running the antiquated projectors. They were not nearly as good as the projectors in Mackinac College's theater, but they mostly got the job done, albeit with breakdowns several times a week.

Movies ran seven nights a week in summer, at 7:00 and 9:00 pm. If it was a particularly long feature, there was only one showing a night. A movie usually ran two days in a row, with Thursday being a solo night. The seats in the Orpheum were rickety and some were broken. Everyone has stories of bats flying around past the big screen, almost nightly.

In the summer of 1968, I worked at my mom and Ray's fudge shop, "Suzanne's Fudge." It was located in part of the shop where Joann's Fudge is now, next to the clothing store. Ray and Jim Dankowski made the fudge. We only carried three flavors, chocolate, vanilla, and penuche, with and without nuts. We also had peanut brittle. Counter clerks were Francis Parker (Jim Dankowski's mom) and an attractive student from NMU, Janine Anderson. In the western half of this building was my mom's upscale gift shop, Voyageur Gifts.

In the fall, came time for 11th Grade. I initially was enrolled in a private school in Prairie du Chien Wisconsin. It was called Campion, and run by the Jesuit Order of priests. It even had an ROTC class for military training.

Campion was strict and all male, although there was a private girl's school just across the river in Iowa. Richard Schwinn attended Campion the same time as I did. He later became the head of Schwinn bicycles. I didn't like attending school there and only lasted a month. I finished out the semester at LaSalle HS in St. Ignace, even though there was now an 11th Grade on the Island. My mother didn't have a high opinion of Mackinac Island Public Schools' (MIPS) new superintendent.

In December, we went to Florida for the winter, trying out Sarasota first. Its high school with 1,600 students was culture shock for me, and thankfully we settled in Clearwater, where I attended Clearwater Catholic Central (CCC) High, which had 300 students. It had a beautiful new domed gymnasium, one of the nicest in the state, which was separate from the classroom building. It was funded, no doubt, from generous parishioners, and the diocese.

We had a dress code of sorts, boys wearing pastel shirts and ties with nice slacks. The girls wore pastel shirts and navy blue skirts. There were all lay teachers, not priests or nuns, but the two administrators were priests, Father Aiden Foines, and Father Jim Finnegan. There was a religion class. I liked CCC, but as with all transfer students, I was not highly popular, but I did have a few friends, and my two cousins attended the school. I finished out 11th Grade at Clearwater Catholic Central, and in May the family returned to Mackinac Island.

In September, we moved down to Florida again, and I went back to CCC for 12th Grade, my senior year. I was asked to be the sideline statistician for the football team's games. Our uniforms were maroon jerseys, gold pants, and all black helmets. I still remember the cheerleaders singing the fight song: "Thunder, thunder, thunderation. We're the Marauders' delegation. We create a big sensation, when we fight with determination." I was also on CCC's golf team. In early winter, we had to take the "Florida 12th Grade Placement Test," similar to SAT's. I finished 10 out of a class of 68 seniors. I'm sure I did poorly on the religion section. Results were posted on the bulletin board, which I didn't agree with. It must have been embarrassing for those finishing 65 through 68.

In mid-March, the family moved back to Mackinac Island. My mom wanted me to finish high school on the Island and graduate there. The class of 1970 was the first to complete 12th Grade and graduate at Mackinac's new school. My classmates were Gloria St. Onge, Jacky Bodwin Bradley, Rose Gallagher, John St. Onge, Jim Bradley, and Lyle C. Horn, who was valedictorian. I was not eligible for valedictorian, even with a 3.85 GPA, since I hadn't attended all four years on the Island. The ceremony was held in the (small) gym, and attended by many Islanders. We all wore royal blue gowns and caps. Superintendent/teacher was Otto Witt, and other teachers that year were Dorothy Hart, Bob Barnes and Jean Bunce. President of the school board was John Bloswick Sr.

In September, I left for Petoskey, having been accepted to North Central Michigan College. I enjoyed my two years there and graduated in late May 1972. When I was not involved in studies, my favorite pastimes were riding my Peugeot 10-speed in the residential areas in September and May. I rode through the streets of Bay View, when all the summer cottages were closed. I also rode in a 50-mile bike race to Wolverine and back, riding solo and finished second against several teams, finishing just ahead of a team of exchange students from Thailand. I also enjoyed being in the bowling league. My team with Ned Taylor, Ross Collins, Frank Solle and the late Tom Gilbert did well. The end of the year banquet was held at The Pier restaurant in Harbor Springs.

Following the Ferries

I have always been fascinated by ships and boats since I was old enough to know what they were. In my early years, through 1966, we lived near the waterfront on the east end of town, making it easy for me to observe and follow them. This included the Great Lakes freighters as well as the local ferries and large steamships. When I was eight to thirteen years old, I made many pencil sketches of many of the older freighters.

The small house I lived in next to Bay View cottage from 1953-57 was built by my dad. He and Matt Yoder also built a small dock there in the late 1940s, made of three wooden cribs filled with stones, and planking for a deck. They were helped in this project by Cowell's yellow tractor. Matt's family owned Bay View next door. The dock had a fuel pump. Dick Welch of Mackinaw City had an arrangement to use it for his 56-ft ferry, "Buddy L." to haul tourists and Islanders from Mackinaw. He was the main competition to Arnold Transit.

The fast mahogany cruisers which used to dock on the west end of town, between the Coal Dock and Iroquois Hotel had mostly gone out of business by then. Their heyday was the late 1930s and 1940s. The last one left was the 48-ft *Fairly Isle*, formerly the *Florence K.*, and she moored at the head of the Arnold Dock.

Not much later, Dick Welch bought three new, single deck, steel fish-tug style ferries from J.W. Nolan of Erie, Pennsylvania: *Christina Mae* in 1954, and *Lance II* and *Mackinaw Clipper* in 1955. They were 63 ft, and had a passenger capacity of 90 each, with limited storage for luggage on top of the engine access platform. They took about 45 minutes to make the trip from the City. *Christina Mae* soon transferred to the St. Ignace run, sailing in tandem with Jack Barnhill's *Treasure Islander*, which was the same size and passenger capacity.

While my mom and dad were married, we rode with Dick Welch. One time while on *Mackinaw Clipper* on a foggy day, we almost got hit by Arnold's *Huron*. It gave me a severe scare. Neither vessel had radar. In 1958, Welch merged with the new company, Straits Transit Inc., made up of ex-employees of the State Ferries, who were from Cheboygan. Their contribution to the new company was the smallest of the ex-State Ferries, the 200 foot, 1000 passenger *The Straits of Mackinac,* which they had purchased. The Island dock was substantially lengthened to accommodate the big vessel, now the largest on the Island run. Tom Bushman of Carp Lake was hired to run the Island dock. There had always been a boathouse at Bay View, which was used for a passenger waiting room. The large *The Straits of Mackinac* was the only Island ferry to have radar in the late 1950s.

In Mackinaw City, the small boats used Dick Welch's dock. He had a small lift for pulling them out in winter. The *Straits of Mackinac* used to tie up at the large former State Ferries Dock, whose ramps were now leased seasonally to Straits Transit and Arnold Transit. The more lucrative slip was the side, east facing one, as it offered protection from the seas. Straits Transit was frequently outbid for it by Arnold, but not always. In winter, The *Straits of Mackinac* harbored at Cheboygan. Strangely, Arnold Transit, which had been in business since 1878, didn't own a dock in Mackinaw City until 1978. They would be tied up at the Michigan Northern Railroad dock, 1878-1957, which was also used by the railroad ferry *Chief Wawatam*. The big new slip at the State Dock, which had been built specifically for the new car-ferry *Vacationland* in 1952, was not used after the Bridge opened.

Arnold had been in business the longest, and in the 1950s was undergoing a modernization of their fleet from coal-burning to diesel power. Long time head of Arnold, Otto Lang, and manager Hugh Rudolph oversaw this transition. The first of Arnold's oil-fired ships was *Ottawa* (2). It was old, built in 1914, but converted to oil in 1926. She served Arnold a long time, often on the Les Cheneaux run, and was replaced by a newer *Ottawa* in 1959. Coal-burning vessels due for replacement were the 140-foot *Algomah II*, which served Arnold Transportation Company from 1936-61, *Chippewa* (3), Leroy Allers ex Beaver Island boat *Mary Margaret*, 1948-61, and *Mackinac Islander*, 1938-57, which had been converted to oil in 1951.

Another older vessel that was sold was *Iroquois*, the former *Detroit*. She served Arnold Transport from 1954-62. Lang contracted with three companies out of Toledo, Ohio; Marinette, Wisconsin; and Erie, Pennsylvania. to build six similarly styled new diesel vessels between 1955 and 1962. Three had cabins and three of them also had an open upper deck. He also purchased the old *Emerald Isle* from Beaver Island in 1963. She would be the only radar-equipped Arnold boat for several years. *Emerald Isle* stayed with Arnold until 1985. All six new builds were lengthened 20 feet from 1965-73.

Captain William Shepler had a speedboat service out of Mackinac City in the 1950s and 60s also. When I was a kid, he had the 30-foot *Miss Margy* and 36-foot *Billy Dick*, mahogany, gas powered vessels he'd built himself. They had white hulls and aqua cabins and docked at the Straits Transit dock.

Captain William "Cap" Shepler got his start in 1945, and his father had piloted Arnold's *Algomah II* for a time. Cap's son, Bill Jr., also worked on the small cruisers. In those days he was known as "Young Bill." They also did night charters and bridge tours. There was no night boat per-se in the 1950s and 60s.

Cap Shepler had two small varnished mahogany cruisers before my time, the *Fiji* and the *Miss Penny*, which he sold in the 1960s. By then he was seeing the need for a larger, double deck vessel if he wanted to grow his boat line. In 1966, he contracted with T.D. Vinette of Escanaba to build him a 120-passenger ferry. The marine architect who designed it didn't have experience with ferries, and the vessel ended up with several design flaws. *Mein Kapitan* only looked good on paper.

I remember Cap showing me blueprints of her, one trip on the *Miss Margy*. The ferry was blue, steel-hulled, double deck, and 60 feet long. However, it had only 15 feet of beam. Her Cummins engines didn't turn out high rpm either. Fifteen feet was simply not wide enough for a double deck ferry, and steel was not the right choice hull for a vessel intended to be fast. Shepler's took delivery in August 1967. Being tippy and top-heavy, ballast eventually had to be added in her hold. Cap and Bill made the best of it for three years. In her first off-season, *Mein Kapitan* was sent to the shipyard in Sturgeon Bay, Wisconsin for modifications.

In 1966, Shepler's also had a new dock built downtown, next to the old State Liquor Store. It was small at first, but was enlarged over the years. Cap's boats no longer used the Straits Transit Dock. In 1969, Cap Shepler took delivery of a new vessel, *The Welcome*, built in Louisiana by an experienced yard, CamCraft. She ushered in the modern Shepler's concept of ferry boats. *The Welcome's* silhouette looked roughly like *Mein Kapitan's*, but the big difference was, she was made of aluminum, with high rpm Caterpillar engines.

She was 60feet long but had a beam of 16'9", 21 inches wider than *Mein Kapitan*. Cap Shepler piloted her up from Louisiana himself. She carried 120 passengers, and even had a few seats on the stern luggage deck, but this was soon removed. *The Welcome* was a total success. She was followed by the 65-foot *Felicity* in 1972 and the 65-foot *The Hope* in 1975. Shepler's had an even larger 77-foot boat, *Wyandot*, build in 1978. All were built in Louisiana. His fleet was very stable, as the main level seating was down a few steps, partially recessed into the hull.

Straits Transit was growing, too, in the 1960s. They briefly leased the *Patricia*, about the size of the Nolan boats, and in 1964-65 leased the double deck, 240-passenger *Islander,* which looked roughly like one of the Arnold cabin vessels. She proved too slow for her Mackinaw City route, however, and in 1966 Straits contracted with Blount Marine of Warren, R.I. to build a 65-foot by x27-foot twin deck ferry, *Island Queen*. She was steel, 350-passenger, had an open upper deck, and was powered by GM engines.

The big steamer *The Straits of Mackinac's* last season was1968, and her last trip, 8:30 pm on Labor Day. Many Islanders showed up on the dock to watch her depart, including myself. Her large crew was aging, and new government regulations were eliminating passenger vessels with wood cabins. This was the same reason *South American* had ceased operation, in 1967.

In 1969, Straits Transit had another Blount Marine boat built, the steel 500-passenger *Straits of Mackinac II*. She was 90 feet by 27 feet with an open upper deck, and twin v12 GM (Detroit Diesel) engines. Her captain for one season was Louis Robarge, who had been the last skipper of *The Straits of Mackinac*, and the purser was ex-Cheboygan hardware salesman Leo "Tut" Louisignau.

Both *Island Queen* and the *Straits II* had radar installed in 1970. Capt. Robarge retired in 1970, after many years of State Ferry and Straits Transit service. He had also been President of Straits Transit since Capt. Emile Potvin's death in 1967. The *Straits II* proved to be a fast vessel, for a displacement hull, and it wasn't unusual for Tut to make a 29-minute trip from Mackinaw City.

In 1972, Straits Transit decided to enter the fast ferry market, and ordered *Island Princess* for 1973 delivery. She would be built in Bay City, Michigan by DeFoe, who did not have much prior experience with that type. She was of aluminum construction, 72 feet by 20 feet, with twin 12v71 Detroit Diesels, radar equipped, and double deck. The *Princess* originally had seating for 250, but due to stability issues, her

Straits of Mackinac II at Straits Transit dock (1970)

rating was reduced to 175 passengers. She had several problems her first year, 1973, and made a couple of trips back to Bay City to have them ironed out, including the installation of larger propeller shafts.

Her captain for her career with Straits was Mike Frey, Straits Transit's youngest skipper. The *Princess* was fast when running properly, and could keep up with or even pull ahead of, Shepler's *Felicity*. She was loud, and suffered from excess vibration. It didn't help that her metal main deck benches were bolted directly to the deck, instead of with rubber gaskets for cushion.

At any rate, Straits now had a modern three-vessel fleet for the Mackinaw City route. *Christina Mae* continued on the St. Ignace run with Jack Barnhill's *Treasure Islander*. *Lance II* and *Mackinaw Clipper* were kept as backup and eventually sold. Other captains working for Straits in the late 60s and early 70s included. Leonard Nau, Paul Repath and Louie Peets.

Arnold Transit equipped all six of their vessels with radar in the mid-1970s. On the foggy days before radar, all the ferry lines used to station a crew member on the bow to listen and keep watch as best as he could, but of course fog played tricks with sound. I remember when I was a kid, Arnold had a fog siren at the end of their 700-ft Main Dock on the Island.

Harry Ryba briefly entered the ferry business in the late 1960s. He ordered two boats from TD Vinette Co. of Escanaba, Michigan. *Ethel Marie* (1968) and *Patricia R.* (1969) were named for Harry's wife and daughter, respectively. They were 61 foot by 18 foot single-deck and painted white. They had a speed of about 22 mph and were loud. The line was called Lakeview Transit, and Harry built a dock across from the Lakeview Hotel. His Mackinaw City dock was next to Shepler's, closer to the Bridge. Locals didn't particularly like riding the vessels. They didn't serve St. Ignace. Harry's line lasted three years, going out of business at the end of 1971. In 1972 Shepler's purchased *Ethel Marie*. *Patricia R.* was sold out of the area. In 1979, Harry's dock on the Island became Star Line's dock.

Straits Transit was out of business by 1978, the victim of a hostile takeover by Arnold Transit, who had been buying up shares for years, until they acquired 51%. Straits had wanted to sell, some of their management was aging, but not to arch-rival Arnold Transit Company. Four captains transferred to Arnold; Tut Louisignau, Paul Repath, David Bohn and Mike Frey, who soon would become Mackinaw City dock manager. The rest were let go.

Former Straits Transit stockholder, Frank Proctor, had gone over to Arnold a few years before, and in 1971 was captain of the *Mohawk*. Arnold initially kept three vessels, *Straits of Mackinac II*, *Island Queen*, and *Island Princess*. The rest were sold. Arnold now owned their own dock in Mackinaw City. *Island Queen* was sold in 1986. *Island Princess* was sold in May 1988, to the Apostle Islands.

Star Line was created in 1979. Island businessman, Tom Pfeiffelmann, was the driving force behind it. Tom had been interested in starting a boat line since 1977, when he, Frank Nephew, and East Bluff cottager John Drudi had an unsuccessful effort to purchase Straits Transit. Several stockholders were recruited, and Sam McIntire was named President of Star Line. Other management included Gene Elmer and Dave Ramsey. In St. Ignace Star would acquire Jack Barnhill's Treasure Island dock, and on the Island, Tom obtained Harry Ryba's Lakeview dock. Initially, they didn't serve Mackinaw City.

Star purchased two old open-bow Nolan boats, the *Flamingo* from the Duluth area, which was renamed *Nicolet*, and the *ex-Patricia/Sunshine City*, which became *LaSalle*. Tom felt they needed a new, fast ferry as well, so they ordered *Marquette* (1) in 1979, from Gulf Craft of Patterson, La. Marquette was delivered via the Gulf of Mexico and Atlantic inter-coastal waterway. She was 65 feet by 22 feet. Veteran captains Red Smith and Harold LaChapelle were her first pilots. Star Line had an ongoing relationship with Gulf Craft for decades.

In 1983, Star Line took delivery of a new *LaSalle* from Gulf Craft, a close copy of *Marquette*, but a bit faster. *Marquette* had bottomed out on a sandbar at Cape May, N.J. on her delivery trip, and had never been 100%. The old *LaSalle* was sold to Leland for the Manitou Island run and renamed *Mishe-Mokwa*. In 1985, Star bought their third vessel from Gulf Craft, *Nicolet (2)*. She had two feet less beam then the previous boats, and was Star's fastest. The old *Nicolet* was sold. Gulf Craft used varnished South American cedar for the trim in pilot houses and elsewhere.

In 1986, Shepler's had a new vessel built, the *Captain. Shepler*. She was the last one built for them in Louisiana, and the last one Cap commanded. *Captain Shepler* was one foot longer and one foot wider than *Wyandot*. They both carried 265 passengers. The *Captain* also had a little more freeboard.

By 1986, Arnold was also ready to get into the fast ferry business. They had been using *Island Princess* as a night boat and overflow vessel out of Mackinaw City. Arnold had a slogan in those days, "We'll never leave you waiting on the dock!" It was a reference to their large, wide-body, 600-passenger Paasch built boats, *Ottawa*, *Algomah* and *Chippewa*.

I recall one busy summer day. *Algomah* had completely filled up in Mackinaw City, and headed for the Island. However, there were still passengers on the dock, so Mike Frey put 100 more on Island Princess, and took them to the Island ahead of *Algomah*. Upon arrival he stepped out of the pilot house and shouted, "We'll never leave you waiting on the dock!"

In 1986, Arnold ordered a state-of-the-art triple deck catamaran from Gladding-Hearn of Somerset, Massachusetts. *Mackinac Express* was 82 feet by 28.5 feet, had a capacity of 350 passengers, and Caterpillar engines with exhaust out the sides of the pontoons. Her speed was 35mph. There was a delay in her delivery, and she didn't arrive until the 1987 season. Arnold used *Island Princess* to keep the already-printed fast ferry schedule they had made. In the 1980s Bob Brown had become manager of Arnold Line.

1988 saw two new ferries arrive at Mackinac. Arnold bought a second catamaran from Gladding-Hearn, *Island Express*. She was a close copy of the prior cat, but with stern exhaust and capacity for thirty more people on deck three. Star Line bought a new hybrid drive, triple deck boat from Gulf Craft, *Radisson*. She had four V-12 engines, two driving props and two driving medium sized waterjets from North American Waterjet Company. She had a capacity of 350 passengers, and was the first four-engine ferry in the Straits. *Radisson*, like *LaSalle* and *Nicolet*, arrived via the Mississippi River and Lake Michigan.

Also in 1988, Star Line added Mackinaw City service for the first time. They built a new dock one hundred yards from the Straits Transit (now Arnold) dock. Shepler's had branched out into St. Ignace in 1986, acquiring the old MRA dock and warehouse at the north end of Moran Bay.

In 1990, Star Line bought their fourth Gulf Craft vessel, the 65-foot *Cadillac*, the first vessel equipped at the yard with their trademark roostertail waterjet, aimed at an angle of 45 degrees, and powered by a smaller six-cylinder center engine.

Starting in the 1990 season, *Marquette*, *LaSalle* and *Nicolet* were retrofitted with center waterjets. *Radisson* had one of her jets turned up at the 45-degree angle also. Since it was powered by a v12 engine, it created a much bigger roostertail.

Captain William Shepler Sr. passed away in May of 1988. It was a sad day for the entire Straits of Mackinac maritime world.

In 1993, Star Line purchased another Gulf Craft vessel, the 65-foot *Joliet*, a close copy of *Cadillac*, except with a smaller pilot house and a higher hull to accommodate Cummins engines. The Cummins experiment did not work out, they didn't turn out the rpms, and she was only a 24-mph boat, so the familiar V12 Detroit Diesels were installed.

In 1996, Arnold Transit bought a third catamaran, this time from Marinette Marine, in Wisconsin. She had a jet drive, sized 101 feet by 31 feet with capacity for 425 passengers and bow seating. *Straits Express* became the largest fast ferry in the Straits, with a peak speed of 40 mph.

All three boat lines made adjustments in the early 2000s to factor in more tourists bringing bikes to Mackinac Island. Arnold lengthened *Mackinac Express* ten feet, (against the advice of Gladding-Hearn). Shepler's lengthened *The Hope* (18 ft), *Wyandot* (7 ft) and *Captain Shepler* (6 ft).

Shepler's *Wyandot* (2017). Photo by Tom Chambers

In 2005, Star Line's last Gulf Craft build, *Marquette II*, had a large rear deck area, for bikes and more carts. *Marquette II* was 80 by 24 feetand featured a state of the art pilot house. She was styled similarly to *Radisson*, but with a larger third deck and lower bow. She was powered by two v16 outboard engines and a small center waterjet. She carried 330 people, twenty less than *Radisson*'s maximum.

In 2015, Shepler's ordered their largest ferry yet, this time built locally by Moran Iron Works, Onaway Michigan. *Miss Margy* was 85 by 22 feet, with three v16 engines, and a passenger capacity of 281. She had a top speed of over 36 mph. Shepler's final boat as of this writing was the 2020-built *William Richard*, 84 feet by 20 feet, a capacity of 220 people and four eight-cylinder waterjets, Shepler's first venture into jet drive. *William Richard* was likewise built in Onaway, Michigan. It's possible Shepler's will have the demand for another new vessel in the next couple of years.

In 2010, the Brown family of St. Ignace, who had owned Arnold Transit for decades, was ready to sell. The buyer was James Wynn, a Petoskey investor. Wynn didn't know much about running a ferry line, and the company went into decline. During their last couple years of operation, they restored good service, crew morale, and professionalism under the leadership of manager Veronica Dobrowolski. She was an ex-Shepler's captain and knew the business well.

After Arnold's last day of service November 10, 2016, Veronica split off the freight end of the company into "Arnold Freight Co," using the vessel *Corsair*, and docking at the Coal Dock for their Island loading and unloading. Shepler's got into the freight business for a few years, with the purchase of a powered barge from Put-In-Bay they renamed *Sacre Bleu* from one of Cap Shepler's favorite exclamations. They made many upgrades on her, with a new, large 55-passenger cabin. Her captain was Greg Bawol. They operated through the 2010s, and then sold the vessel to Arnold Freight, giving ATF a much needed second vessel.

Through much of my lifetime, Arnold Transit's senior captain was Paul Allers. When he was young, he got his start on the speed cruiser *Fairy Isle*, and also Arnold's old *Ottawa*. He had a stint in the US Navy, then returned to Arnold where he worked year around, and was known for piloting the winter boats *Mackinac Islander* and *Huron*. Paul had decades of service as a captain and became the most respected in the Straits. His dad, Leroy, had worked for Arnold, also.

Other winter captains were Ray Wilkins and Lloyd Hulett. In the late 1980s, Arnold experimented with naming a "Fleet Captain," Who was Jim Brown. Star Line's senior captains I remember were Mark Brown, no relation to the St. Ignace Browns who came over from Shepler's in 1979, and Larry Spencer. For Shepler's, their senior captains (not named Shepler) were David Sullivan and Mark Inglis.

Star Line purchased Arnold Transit, and their valuable docks and St. Ignace Mill Slip, in the off-season of 2016-17.

Some of the good boat-handlers of the later 20th Century have passed away now, Tut Louisignau in Florida, March 10, 2008, Paul Repath on Nov. 23, 2017, David Sullivan on Dec. 11, 2021, and Mark W. Brown on Feb. 23, 2022. Also, Jeff Davenport of Star Line, who was younger than me, passed away on Jan. 5, 2017. He ran the *Nicolet* out of the St. Ignace RR Dock. Captains in the Straits who carry on the tradition of good seamanship lately are Preston Allers, Sean Whelan, Doug Derhammer, Sid Hawkins, Jim Ryerse, Geoff Hoeksema, Steven Buchannan, Kyle Wallis and Justin Davenport.

Each of the boat lines had their share of "incidents," as the captains liked to call them. In the days before radar, they usually consisted of running aground in the fog, or grazing the west breakwall or the rocky point of Round Island. When I was about five years old, I was playing in our backyard, next to Ste Anne's rectory, and there, not 200 feet off shore, in heavy fog, was Arnold old *Chippewa*, stopped and with a list. I went running in the house and exclaimed, "Mom, the *Chippewa* is in our back yard!" I also remember stories of Arnold's old *Mackinac Islander* colliding with an obstruction.

Over the decades, several vessels had a steering cable break, and had to be towed by a fleetmate. *Huron* even ran aground off Round Island, 6/8/63, and 20 passengers were removed. One time in about 1979-80, *Emerald Isle*'s complicated gear shift failed to respond from reverse, and she backed into a yacht at the yacht dock, much to Capt. Jim Smith's chagrin.

On her first day of service in 1988, Star Line *Radisson's* broke a jet pump about a mile out of Mackinac City. The boat started taking on water, and Captain Mark Brown turned her around and got her back to the mainland dock to tie up. She carried on until October with only one of the inboard jets operative, and went to The Soo to have it repaired in October.

One time in her opening season of 1996, Arnold's catamaran *Straits Express* experienced computer failure while docking at the Island. The helm would not respond, and she coasted into Ryba's Bike shop next to the Chippewa Hotel. Marshall Brother's construction was there within the day to repair Ryba's bike shed.

In around 1991, Shepler's *Captain Shepler* lost power in the channel on a day with a strong west wind, and drifted into the east breakwall. The story I heard was that she had lost her gas cap while refueling her diesel fuel, and they jury-rigged the opening with duct tape. At any rate, Shepler's quickly sent *The Welcome* from their Island dock to get a line on the Captain, and try to pull her to safety, but *The Welcome* didn't have enough power.

When *Wyandot* arrived in the harbor from the mainland, she towed *Captain Shepler* in, where she remained docked at the end of the Coal Dock for the rest of the day. I wish I had gotten a photo of this one-time mooring.

In the early 2020s, Star Line's *LaSalle* failed to answer the helm, and hit the beach at their St. Ignace Railroad Dock. In summer of 2023, Star's Mackinac Express lost power in the channel, and soon was towed to the Island by *Radisson*. These are just the incidents I remember. I'm sure there were others.

In the early 2010s, Tom Pfeiffelmann retired as General Manager of Star Line. His wife, Linda, succeeded him for a couple years. After that, the head position should have gone to popular and capable longtime senior captain Mike North, but the stockholders chose to go in another direction.

Traveling By Air

My first memories of airplanes at Mackinac were watching them land and take off from the harbor ice in the mid-1950s. Ice in Haldimand Bay was good almost every winter in those days. Several Islander locals learned to fly, and had small 4-passenger aircraft, usually high wing. Planes gave them another travel option. Skis were mounted from the struts where wheels were used in summer. I remember brothers Jim and Al Francis flying. Also Dennis Brodeur, who sometimes landed on the Grand Hotel golf course in winter. Mackinaw City had a small air strip, but St. Ignace Mackinac County Airport was the most popular.

There were licensed pilots operating out of St. Ignace, Al Phillips and a Mr. Wartella, who provided passenger service to the general public in the 1950s and 1960s. The Island's airport was pretty primitive during the 1940s to 1960s. The oldest photo I have seen was dated 1934. It was a grass/dirt strip about 2000 feet. long. There were no lights for night travel. The "waiting room" was a small tarpaper shack, around 10 by 12 feet, with a wood stove in the center. The attendant I remember was State Park employee Frank "Yunk Yunk" Visnaw. He would greet Islanders, give flight info, and tend the wood stove. Frank's kids, Steven and Rose, were classmates of mine at the public school.

In 1964-65, the State of Michigan modernized the Island's airport: paved it with asphalt, the length increased to 3500 feet. Paved side areas for airplane parking were included. I remember them saying the new strip could accommodate a Douglas DC-3 size aircraft. A new terminal was built with a waiting room, offices and restrooms. I recall State Park Employee George Bodwin being the new attendant.

A large barn-type storage building was also built in the coming years, off to the side by the wooded entry road. In the late 60s, Mackinac County's pilot that I remember was Bob Fennell. I went to school at LaSalle in St. Ignace briefly with his daughter, Deanna. In the 1970s, a popular pilot with Islanders was John Geisinger, flying a Piper Cherokee six-seater.

By the 1980s, even small jets landed here, mainly the Lear-Jet. Lighting was added over the years for night service. Islander Joe Kompsi took up flying, and he and Arnold Line manager, Bobby Brown, each had small high wing craft they often parked on the ice near the Arnold Dock. In winter they also scouted ice conditions on the route for Arnold boats *Mackinac Islander* and *Huron*.

Around the turn of the century, off-island aerial interests wanted to enlarge the airstrip further, across British Landing Road, to accommodate larger jets. The Islanders were alarmed by this plan, held meetings, and firmly came out against the idea. The plan was scrapped, but the airport was leveled in some areas, and new up-to-date lighting added.

Mackinac Island Airport (1966)
Photo by Don & Lucy Gridley, Alan Bennett – cab driver (inset)

When 10-Speeds Came to Mackinac Island

One thing about living on Mackinac Island is, you probably owned a bicycle. I went through your typical series of kids' bikes, and then graduated to a new Sky-Blue Schwinn Corvette 3-speed middleweight in 1963. It had hand brakes and the popular Sturmey-Archer 3-speed shift on the right handlebar. I saw my first 10-speed bike in around 1963, and became fascinated by them. They operated on a different principle than 3-speed, known as a derailleur. They had skinnier tires, and 1 inch larger diameter wheels.

The first one I saw was a red Schwinn Continental, owned by May's fudge maker, Carl Lake. Carl had put Stingray handlebars on it, as some found the deeply curved "drop bars" used on 10-speeds to be uncomfortable. Carl was a cousin of Marvin May. He soon moved up to an imported French "Louison Bobet" 15-speed, teal in color, also with Stingray bars. He sold his Continental to Island teen Dorman Kompsi (nephew to Joe). One time, a bunch of us kids gathered around Dorman and his bike, down by Mission Point. "You kids couldn't even pedal this thing in tenth gear!" he ribbed us.

Another early bike I saw was a 1964 Coppertone Schwinn Varsity, owned by Bob, the Borden's Milk delivery man. Varsity was the entry level Schwinn 10-speed, the next model down from Continental. Annex teenager, John Croghan Jr., had a nice Italian Frejus. The first of my Island friends who got a 10-speed was Billy Fallstich, with his 1964 Sky Blue Varsity. By 1965 I decided I had to have one, and ordered a Sky Blue Schwinn Varsity Tourist from Island Bicycle Livery. It cost $69.95. The Tourist model had standard handlebars, a mattress saddle (instead of the leather racing one) and chrome fenders.

I loved it. I kept it one year, and then moved up to a 1966 Schwinn Super Sport, in Coppertone. It had the slightly lighter chrome-moly frame (as opposed to steel), drop bars, leather racing saddle and chrome fork. I should mention, bike rental outlets on Mackinac never chose to offer 10-speeds in the early decades. In summer 1967, while living at British Landing, I met a guy a couple years older than me, named Skip Hill. He was visiting the Island, and had a true racing 10-Speed, a white French Peugeot PX-10, their top of the line. It weighed only 20 lbs., which was extremely light for that time. I was mesmerized by it and vowed to have one someday.

I purchased one when I was going to school in Clearwater, Florida, 1969, and paid $150 for it. I brought it back to the Island from Florida. It served me well for four years, at Silver Birches, working at the Village Inn, and at North Central Michigan College (NCMC) in Petoskey. I entered it in a 50-mile bike race from Petoskey to Wolverine and back, in May 1972. I rode solo, competing against five 5-man relay teams, and took second place. I originally had a team, but they all dropped out. It was a hot Sunday afternoon, and there was a long, steep hill about six miles out of Petoskey on the return trip. I ended up having to get off and jog my bike up the hill, since I had no true low gears. I saw a guy sitting on his porch, and I bummed a glass of water from him, and sat and talked. After a few minutes, I noticed a rider from the Thai Nationals team coming up the hill in the distance. I said, "Well, I better get going," got on the Peugeot, and finished the final leg back to NCMC.

In spring of 1973, I ordered a black Schwinn Paramount P-13 road racing bike from Island Bicycle Livery. Tom Pfeiffelmann had been bugging me for years, trying to get me on a Paramount. The cost was $450, which raised eyebrows all over town. My cousin, Ron Dufina, even commented, "You've got to be nuts to pay $450 for a bike!" Years later, he paid over $1000 for a Cannondale mountain bike. I was not the first Paramount owner on the Island, however. That honor went to Stewart Woodfill, although his was a 3-speed tourist model. My Paramount was deluxe all the way, with fancy chrome lugs. At 23.5 lbs, it was a bit heavy for a road racer, 3.5 lbs heavier than my Peugeot.

That summer was my first year living at my grandpa's old cabin at British Landing, and I rode the Paramount every day to my job with the City. I raced it once that summer, in an individuals round the Island race. I didn't place. I discovered the 25-inch frame I ordered was too big for me, so before left for Western Michigan University (WMU) in the fall, I sold the bike to Vera Francis. Over the winter I bought an Opaque Green P-13 Paramount from my roommate, David Phinney, as he owned two. It had custom pin-striping and a 24-inch frame, which was just right. It came back to the Island with me for the 1974 season.

I raced it several years in the team relay races. It was repainted black around 1980, and then a light metallic blue. I still have the bike to this day.

All of my racing bikes, starting in 1969, had tubular "sew-up" tires, which were skinnier, and glued directly to the rim. They really weren't practical for Mackinac Island, but they gave a bit of an edge for racing compared to the conventional tires of the 1970s. For the relay races, Tim Horn and I used to take our Paramounts up to the State Park garage, and inflate the sew-ups to 110 lbs pressure.

When Snowmobiles Came To Mackinac Island

Every few decades a major change takes place on Mackinac Island. Snowmobiles were one such change. Personal motor vehicles had been banned on the Island since 1898, which contribute to the Island's allure and charm for tourists and locals alike. It also fostered the supremacy of horse-drawn transport on the Island. Commander E.M. Tellefson had put this law to the test in the 1930, when he tried running a car from his Pointe Aux Pins property to his radio shack and tower at Fort Holmes, but he lost his bid for it to be allowed.

The forerunner of the snowmobile was the motor sleigh, or "air sled." These consisted of a wooden body about 15 ft long with three skis, one forward and two in the rear. An open propeller-driven airplane engine was mounted on the rear. These were popular in the 1940s through 1960s and operated on the ice, but almost never on a street. They carried a driver, a couple passengers, and a small amount of freight or groceries. A few ice-experienced Islanders owned and operated them: John Bloswick, Ray O'Brien, Dean Gillespie, Jim Perraut, and brothers Dale and John R. "JK" Gallagher.

Primitive snowmobile-type vehicles had been around since World War II military use, and around 1960 they made the jump to recreational vehicle. It goes without saying that they became popular with Americans. In the period roughly 1963-66, Mackinac Islanders were split roughly 50-50 regarding this new vehicle.

The first group was adamantly opposed to them. The grounds were, snowmobiles were a precedent-setting violation of the "no motor vehicle" ordinance, and if they were allowed it would become a slippery slope. Additionally, they argued, the machines were noisy and a source of pollution, fouling Mackinac's pristine, clean air. Advocates of snowmobiles saw then as a godsend for winter existence, both for transportation and for hauling goods and personal supplies. Not every resident had a barn, or the wherewithal to drive a team of horses, or a horse and sleigh. Snowmobiles would be a practical alternative.

Snowmobiling on the Ice (1988)

In 1963, Elmer "Bud" Bradley, whose family lived at the power house on the shore road past Arch Rock, brought a red Arctic Cat to the Island. He wanted to use it to ride his four kids to the Island school, and to haul supplies to the out-of-town power house. The machine was slow, about 10 mph, mostly made of steel, and very basic. Bud didn't use the roads, but traveled the shoreline ice and across the harbor ice. Others soon got the snowmobile bug. I think Tommy Bunker in Harrisonville and Jim Francis at the East end of town were next, followed by Bud Emmons, Ty Horn and Armand "Smi" Horn.

Most bought the Polaris Colt or Mustang models, 12 and 15hp respectively. These sleds had a white hood, red seat and pale blue chassis and belly pan. My family moved to British Landing in 1966, to the old Helen Wagner cottage. We had beachfront property, and my stepdad, Ray Summerfield, bought a Polaris Colt, then upgraded to the larger Mustang, for hauling us kids to school on the shore ice. So, by 1966 there were seven snowmobiles on Mackinac. My mom, however, continued to remain opposed to snowmobiles.

The snowmobile industry was really taking off nationwide, and by 1969 there were probably 100 brands. Anyone who made an outboard boat motor and some who made farm equipment was making a snowmobile. Even Sears and Montgomery Ward's had them. Engine size and features were improving also. Their numbers grew on the Island, to a couple dozen by 1969.

A quandary arose, both with the State Park and the City, regarding what to do about them. The State was rigidly opposed, the City was lukewarm. The path to snowmobile legality came in small steps at a time. No entity cared if you rode one on the ice. The next step was to allow them on private property. In 1968-69, there were a couple dozen sleds on the Island. Legal road usage was still out of the question. Islanders developed novel ways to get from Point "A" to Point "B". For instance, villagers would cut through the Big Barns, and along the Grand Hotel Golf course. They would park at Little Stone Church or the schoolyard, or go around the end of the boardwalk and into town at Windermere Point. If you crossed a road along your trip, the authorities would look the other way.

The ice between the Arnold Dock and Coal Dock was used as parking for Mustang Lounge patrons. High school kids also used to joyride on the Grand golf course. The City was starting to soften on snowmobile use, but the State Park stood firmly against it. Something had to give. I was not here to see it, but in the winter of 1969 or 70, a protest ride was organized. Their route was through Main Street to the east end of town, up Mission Hill, and past the State Park offices behind Fort Mackinac. Ringleaders were Bud Emmons, Jack Chambers, Smi Horn and Danny Wightman. A dozen or more sleds took part. This led to The City making snowmobile use on City streets legal. Soon, the State Park also opened up, M-185 and the entire west half of the Island, Garrison Road and British Landing Road included.

Snowmobile travel to St. Ignace began slowly in the late 1960s, mainly by seasoned older Islanders leading the way. The practice of putting Christmas Trees out as trail markers began.

Places like Arch Rock and Fort Holmes were still illegal, but East and West Bluff, the Annex, the Airport and Stonecliffe were allowed.

In 1970, Jack Chambers and Smi Horn opened an Arctic Cat dealership in part of the old Village Inn building. Ron Cowell had a Massey Ferguson dealership and sold them from his house on Spring Street. Over twenty brands in all colors could be seen around town now. The Douds had an Evinrude. Rollie McCready had a Johnson. Jim and Donny Francis had a Mercury. Liquor Store manager Jay Parker had a Rupp. Riding stable manager Tony Lordson had a big 634 Arctic Cat, with chrome side pipes. Dick Bagbey and Jim Bradley had Skiroules. Vin Francis had a Ski-Doo, complete with cigarette lighter. Phil and Armin Porter bought John Deeres. There were Yamahas, Sno-Jets, Kawasaki and Ski Whiz. There was even a "snowmobile club."

In the mid 1970s when Fr. Joseph Polakowski was priest at Ste Anne's, his younger brother Fr. Tony would visit sometimes. He drove a white Yamaha 1974, which had the novelty of a dual-trumpet chrome boat horn mounted on the hood, and through the windshield. In summer, on the mainland he drove a Corvette. I believe he later died in a car accident.

I used to drive my dad's Arctic Cat Panthers until I got my first machine, a 1977 Sno-Jet in 1982. In 1987, I got a top of the line Kawasaki Invader, which I had until 1999, and in around 2015 started riding a 1998 Polaris Indy Trail, which had belonged to my late buddy, Armin Porter.

Tourists didn't cross the ice to the Island in the 60s or 70s. That phenomenon would start in the 1980s, and only slowly. It really took off in the 1990s and with the advent of social media on the internet. It's hard to believe, but all but four snowmobile makers were out of business by about 1984.

First Band, 1969-73

I spent Christmas break 1968 at home on the Island, having finished the semester at LaSalle HS in St. Ignace. During this time I met Matt Finstrom, a guy my age whose parents worked at Mackinac College. I found out Matt played drums, and we arranged to have a practice at my mom's summer shop, Voyageur Gifts, located on the west end of downtown, where Joann's Fudge now is. It went well, and we agreed we should try to put a band together the following summer. My family left for Florida in January, and when we got back in spring, Matt and I reconnected, and started practicing in Ste Anne's Church basement.

My Island friend, Billy Fallstich, joined us. I still had my Kingston electric guitar I got at Woolworth's in Cheboygan, and Billy had a real nice Gibson SG/Les Paul model from around 1961. It would be worth a lot of money these days, but unfortunately Billy had it stolen a few years later on a trip to Marquette. He and I alternated on guitar and bass. Billy also had a nice little Fender Princeton Reverb amp that he paid $100 for. Billy ad I also knew singer, Vince Palermo, two months older than me. His parents owned the wonderful Chatterbox Italian Restaurant downtown for many years.

Billy and Vince used to spend a lot of time in Marquette Park, Vince crooning, and Billy accompanying him on acoustic guitar. They attracted a lot of girls passing by. We asked Vince to sing for the new band, and he came to Ste Anne's for a couple practices, and to learn the set list. This was 1969, and we played current songs, as well as rock classics from earlier in the decade. Songs such as Sitting on the Dock of the Bay and Gloria were popular. After a trial performance in late June at the Community Hall, mostly for friends, we booked another gig there in July. While we were playing, some girls came in who knew Vince, They started talking to him onstage, and he decided to leave with them, right in the middle of the gig. Since Billy was his best friend, he went with them.

Vince had never had band discipline or commitment, but needless to say, they left Matt and me in a fix. Just then our friend and schoolmate, Gayle Welcher, came up to the stage and told me she knew two guys in the audience who played in a band in Detroit. So I told her, "Bring them up!" She introduced us to Don Meadows (bass guitar) and David Nephew (singer), our age, who played in the Dearborn Heights band Orange River. It was their first summer working on the Island, for Frank Nephew. They sat in, finished the gig and bailed Matt and me out.

Afterward, I asked them if they wanted to join the band. They did, and we booked our first official gig at Community Hall in late July. I chose the name "Armageddon," after a Star Trek episode I had seen, called A Taste of Armageddon." A year or so later, concert pianist Richard Hadden, an Island resident, told our drummer Matt there was already a band by that name, so I came up with a new one. It was Morningstar, named for the medieval ball and chain mace. Our logo became a circle with five points around it. I believed every band should have a distinctive logo, ala Blue Oyster Cult's.

Things took off by the spring of 1970, and we played Mackinac Island's HS prom. We played in the Community Hall again for the Lilac Queen's Coronation Ball in June. We had "uniforms" by this time, four different military coats picked up at Mackinac College's costume room. They were wool and hot, so they often didn't survive very late into a gig. The Beatles Sgt. Pepper had influenced us in the decision to get the coats.

We would also pick up a Guild Thunderbass 2x12 amp from Mackinac College. By spring of 1970, I realized I needed a new electric guitar. I went to Joe Hanna's Music store on Lake Street in Petoskey. It was a big store with a full selection of everything. He had used guitars, and an old beat-up sunburst Fender Stratocaster and Vox Hurricane were each priced $100. Vox was an equal brand to Fender back then, and I chose it because it was in better shape. I still wonder what that Strat would be worth today.

In September, we played a gig at North Central Michigan College in Petoskey, upstairs in the student center. I was using a Jordan solid state amp with 2x15 speakers. Don had a nice Wards bass combo amp, only three controls, but with 6x10 speakers. Our PA was a Fender Bassman head with homemade speakers David and Don had put together, with 1x12 and a large horn in each column.

By the early winter of 1971, we realized we needed a lead guitarist. I would continue on as rhythm guitarist and keyboard player. I had bought a Farfisa Compact organ in Florida the previous winter. Matt had moved to Traverse City for the school year and had been practicing with a schoolmate/guitarist named Steve Colbath. Steve was originally from Boyne City, but had also moved to Traverse. So, Matt asked him to join our band. We were now a five-piece outfit. Steve was a quiet, soft-spoken guy and a skilled guitarist, with a beautiful new Gibson Les Paul gold top guitar.

We also were fortunate to have a "band tech" in 1970-71. Bill Henry, a bespectacled 6'5" nerd and recent graduate of Mackinac College, kept all our amps running. I still know Bill to this day.

Our first gig with the new lineup was eventful. We were booked to play a winter dance at Mackinac Island HS in early February 1971. The ice was good that year, so I got my dad and two of his friends, Bud Bourisaw and Ducky Smith, to pick us up with snowmobiles and cutters in St. Ignace. Moderate snowfall was coming down all day. Jim, a friend from NCMC, and I drove up from Petoskey. Matt and Steve, along with friends Pat McGee and Brad Graham, drove up from Traverse City. A side note here—Brad Graham would later go on to have a career as a news anchor on some Michigan TV stations.

Our other band members, Don and David, drove up from Detroit. However, the snow was worse down-state, along with wind, and they got in a car accident near West Branch. They were OK, but the vehicle was no longer drivable. It was decided that the three band members in St. Ignace would have Jack, Bud and Ducky drive us over to the Island with our gear, so we could set up and "hold the fort."

Meanwhile, our other friends who came along would drive down to West Branch and pick up Don and David. After my dad and his friends returned to St. Ignace, they had a few hour wait for the rest of the band, so they naturally spent time at one of the St. Ignace bars. All this was in the age before mobile phones, of course. By the time the crew got back from West Branch, Jack and friends had quite a few drinks. The snow was also coming down hard, and it was getting dark. Buddy was in no shape to drive, so he rode in a cutter and one of our guys drove his sled. They set out across the ice, and through the middle of the Island, a three-sled convoy, with Jack in the lead.

They arrived at the school none the worse for wear, just with speakers full of snow. Matt, Steve and I had played a couple sets of music for the high school crowd and chaperones, while we waited for the others. The night's gig went fine, and the kids were happy. I do remember our bass player, Don's, Island girlfriend breaking up with him that night. Don used two basses throughout his career with us, a Fender Precision and a Fender Jazz, repainted in powder blue.

In May of 1971, we landed a lucrative gig at ski resort Schuss Mountain, in Mancelona. It was for Bellaire High School prom. My NCMC friend and fellow musician, George Blanz, had gotten us the gig when his band was unable to play it. We had a busy night, as the ballroom next to ours also had a prom and band. I had gotten a new amplifier on approval from Ware's Music in Traverse City to try for the night. It was called a Baldwin "Exterminator." It was solid-state, blue and black, huge, with a bunch of colored push buttons on top and six speakers inside.

We also developed a novelty starting with this gig, where a band member sat out a song, and go out into the crowd and listen to the band, checking how the sound was. Mine was our new cover of "Baby I'm Amazed" by Paul McCartney.

I didn't end up keeping the Baldwin amp, it was too heavy to lug around, and was too clean sounding. It didn't have the ability to distort at higher volume.

I did however pick up a new guitar in Traverse City for the Schuss Mountain gig. It was a nice 1965 candy-apple red Mosrite, endorsed by my old guitar heroes The Ventures. These saw their heyday between 1963 and 1968 but were out of favor now. Some guitar brands were trendy, and didn't have the staying power of a Fender or a Gibson. I paid $100 for a guitar that originally sold for $350 new, and had competed with Fender Stratocasters.

I was also using a fuzz tone pedal on some songs by this time, first a Gibson Maestro and later a Mosrite "Fuzz-Rite." A couple of weeks after Schuss Mountain, we played the Park Place Hotel in Traverse City for a Kiwanis Club function. We had to wear sport coats and ties. Steve brought out a 1960s red Gibson SG for this show. His normal guitar for 1971 and most of 1972 was a 1971 Gibson Les Paul Deluxe Goldtop.

Of course, for playing mainland gigs we needed transportation and for hauling the band gear around. Steve's parents had given him their mid-60s blue Chevy full-size station wagon. We also had an olive green early 60s Lincoln with the "suicide" doors. I think Don had gotten it from his dad, Forrest. Our gear plus the five band members, maxed out the two cars.

While we were in town for the Park Place gig, the five of us piled into the Lincoln and took a ride out to Steve's parents' house on the inland side of Peninsula Drive. Steve mentioned to us that Michigan Governor William Milliken also lived on this road. Don, who was driving, asked "Where?" Steve pointed it out, and on impulse Don made a left turn into Milliken's long driveway on the water side of the peninsula. We found the Governor raking leaves in his yard, with his small dog running around. We pulled up, got out, and introduced ourselves. After about a five minute chat, the Governor told us "If I ever need a rock band while I'm on Mackinac island, I'll get in touch with you!"

I needed a new amp, so I bought a big stack from Wards in Petoskey. It had 4x12 speakers and a horn, and casters, making it easier to move around. I didn't really like the Wards head, it was solid state and awfully clean, so I sold it before I left NCMC at the end of May. I went to Joe Hanna's Music on Lake Street and bought a nice Fender Twin Reverb, 85 watt tube amp with 2x12" speakers. It cost $400, and my dad helped me with it. So my "stack" for summer was the Fender Twin with the Wards cab, a total of 6x12s, standing over five feet tall. It was actually too much amp, and more than Steve's 2x12 Silvertone combo. We lined up two gigs at the Community Hall, first the Lilac Queen's Coronation Ball in June, and later the Fireman's Ball in July.

1971 was also our first summer of doing "Booze Cruises" aboard Straits Transit's ferry *Straits of Mackinac II*, evenings in July and August. These were organized by me and Bruce Jorgensen, a bartender, dock-porter, and influencer about town. We sold tickets to friends and the local summer employees. It was bring your own booze. Bruce provided some mixes for the liquor. Tut Louisignau of Straits Transit was the captain for all our cruises and was happy to do it.

After loading at the Straits Dock, the band set up at the starboard gangway. A few rows of seats were removed for a dance floor. *Straits II* was a 500-passenger boat then, so the lower deck was pretty full of seats, except for a little space for two or three luggage carts. During one of the '71 cruises, we passed close to one of the Republic Steel triplet freighters in the South Channel. She was steaming eastbound at 20 mph. Dancers were weaving and losing their balance.

Late in August, before heading back to college, I sold the Fender Twin Reverb to Jim Enrietti of The Zippy Zips. He used it for his electric accordion. We didn't have any gigs planned for fall.

Sometime in October while at NCMC, I had a phone call with bass player Don Meadows. He was the last band member still staying out at Silver Birches. I remember us getting into an argument about something, and I kicked him out of the band. It was a moot point, because Don was in the process of enlisting in the US Air Force, anyway.

There was a sad postscript to Don, our bass player of two and a half years. On New Year's 1979 he was at a dinner party in Detroit, and he became a victim of a homicide, shot with a handgun. I was on a trip to Europe at the time, and when I got back, David Nephew showed me the newspaper article on the 2nd front page of the *Detroit Free Press*, titled "A Murder At Dinner, But Why?" He was 26 years old.

In early winter 1972 it was time to get a new bass player. I recruited a classmate and friend from NCMC, Dave Pagel, from Gaylord. He agreed to join Morningstar, and work the summer of '72 on the Island. Dave was a few months older than the rest of us and the only band member born in 1952. The rest of us had been 1953. Dave was good, learned our songs fast, and had great equipment. His bass was a real Hofner sunburst Club Bass, and his amp a tube powered 100 watt West piggyback. It was made in Michigan by Dave West. It had 2x15 speakers and weighed a ton. West amps were also used by Grand Funk Railroad at this time.

Our first gig with Dave was in Feb. 1972 at The Dillworth Hotel in Boyne City. It was a rare two-night performance for us, and was in the hotel's basement Rathskeller room, which room was good-sized, with peanut shells all over the floor. There was a snowstorm all weekend, and we could not even drive back the short distance from Boyne City to Petoskey after the Friday night performance. Both nights went well, and Dave Pagel proved a great addition to Morningstar.

On May 3rd 1972, we returned to NCMC for another gig, this time in the basement of the Student Center. Steve brought out a beautiful new black Gibson customized 335 hollow body for this gig, and an old Fender Strat he used on a couple songs. Some of my college friends and teachers attended, and the crown of about 100 people liked us. I was still using my red Mosrite guitar, and for keyboards I had Farfisa and

Vox combo organs. I also debuted a new amp in 1972, a Sunn tube head with a 4x12" cab made by Acoustic Control. A couple of weeks after our gig, the Michigan band SRC played NCMC, under their new name Blue Scepter, but still with their classic lineup. At the end of May, Dave Pagel and I graduated from NCMC, and the band headed to Mackinac Island for the summer season. Dave got a bartender job at the new French Outpost bar owned by Sandra and Bob French. They had purchased Mary's Pantry that year and renamed it. In the fall and winter of 1972, Jim Francis built them a new bar, the one everyone remembers. Sandra and Bob had foresight about hiring rock bands to perform, and booked us for early June. We set up near the long row of windows in the back of the bar. We had a sixth member for this gig, a friend of Matt and Steve's from Traverse City named Bill Stebner, who played additional percussion and congas. Don Meadows was home on leave from the Air Force, and even came to hear us. My friend, Billy Fallstich, was there too, and introduced me to a girl I ended up dating for the rest of the year.

We were really loud for that gig, as all of us had big amps. A singer friend of mine from the band at Grand Hotel said her bathroom fixtures were vibrating at Grand Cottage, half a block away. Steve now had the amp he would use for the next two years with Morningstar, a 40watt Fender Super Reverb 4x10, with a 2x12 extension cab of Michigan made SRO speakers. He didn't use effects pedals, but got great sustain and distortion anyway by turning that rig up to 9. Some nights David's PA was hard pressed to match volumes with our amps. I was using a new 1972 cherry sunburst Les Paul Deluxe guitar. It had a phase switch and was rare, because of the block inlay on the first fret.

We had planned on making a reel to reel recording of this gig, as we had done a month before in Petoskey, and it turned out to be one of our best-sounding performances. Unfortunately, the recorder was set up behind our wall of amps, and an open beer vibrated off an amp, landing on the tape deck, and ruining the tape. A couple nights later the girl I met gave me an impromptu haircut under the security lights at the school, since my dad was on my case to have shorter hair for my new bartending job.

Another band got the Lilac Ball gig in '72, but in July we again played the Fireman's Ball. We all bought new, fashionable shirts for the gig, and borrowed a set of risers from The Zippy Zips, so we would have a two-tiered stage. For most of our Community Hall gigs, we were hired by Kay Hoppenrath. We did two booze cruises on the *Straits of Mackinac II* again, with Capt. Tut Louisignau, in July and August. The July one was during Yacht Races, and while we were warming up at the Straits Dock, some yachters came over and asked us to turn down. So when we left, instead of Tut pulling directly out of the harbor, he swung the *Straits II* west, past the yacht dock, and along the Arnold Dock, while we were on our opening song.

Also in August, we played gigs on two consecutive weekends in St. Ignace. It was at a new teen center called the Maxwell House, in the old Walker Furniture Store location on the waterfront. I think it was run by a guy named Rick Engen. We took the 8:00pm *Arnold Chippewa* over, as we didn't want to try to fit all our gear on the *Treasure Islander* or *Christina Mae*, along with tourists. The *Chippewa* was still short that year, not yet having been lengthened 20 feet. David Nephew's longtime girlfriend, Lynne, came along on the second week's gig, and took some photos. At about 1:30 in the morning we took Alfred Thibault's charter yacht back to the Island. He had a nice, varnished mahogany 30 ft cruiser. Some of the guys smoked joints, which made me a bit nervous in front of Capt. Alfred.

I used to go to Grand Hotel on some of my nights off to hear their rock band. The Grand had hired a band called "The Music School" to play alternating sets with their house orchestra that year. It was six-piece, with Joel Balin on a nice Gibson Byrdland guitar, Skip Callahan on bass, Bruce Schuler on keyboards, and Larry Hays on a chrome Ludwig drum kit. The singers were Joel's girlfriend, Trace, and Larry's girlfriend, Betsy. The girls wore evening gowns and the guys wore black tuxedoes. They played mostly soft rock, current hits of the day. One night, Steve came with me and we brought Village Inn servers Cissy Springate and Suzie Hopkins. We became friends with the Grand's band, and Joel and I would sometimes hang out and ride ten-speeds. Steve was normally dating my dad's niece, Maryanne, that summer, but she had gone back to Chicago with her family. Uncle Bill, Aunt Helen and the kids had been renting an apartment from Bud and Betty Emmons.

Labor Day weekend was the last hurrah of the summer, and Carriage Tours driver and colorful character "Handsome" Harry Foster had been selling tickets all summer for his "Pardner's Club." Harry planned a Labor Day night party at the end of the summer for his "Pardners," held at the end of the boardwalk, and Harry provided all the canned beer. He asked my band to play, and we got Milt's Dray Line to provide a dray for a stage. Louis Spencley, the old drummer from St. Ignace band GNP, drummed for us all night, since Matt was at U of M in Ann Arbor for orientation. The audience consisted of islanders, summer employees and business people. Harry sold 430 tickets, I believe. Even Harry Ryba showed up, and my dad. I think it was the only time my dad ever heard us play. I recall him getting in a disagreement with the major domo from Hedgecliffe, a guy named Curtis.

We had guest musicians, too, redcap porter Julian Wright on vocals, and Billy Fallstich and Joel Balin on guitar, which allowed me to go out in the crowd and take some band photos. George Archer, an ex-bartender from the original Village Inn, also took some good black and white photos. The gig went well, and we all had a good time. After Labor Day, it was time for most to leave Mackinac. Matt and Steve had been cooking at the Village Inn, David was working for his Uncle Frank again at Pickle Barrel and The Orpheum, and Dave Pagel and I had been bartending, Dave at French Outpost and me at Village Inn. No fall gigs were planned. I spent fall on the Island, continuing to bartend at the VI. I would leave for college in January, new to Western Michigan University in Kalamazoo. It was my junior year.

Morningstar had no gigs in the winter of 1973. In spring, when my semester at WMU was done, I returned to Mackinac, for a summer job with The City. David Nephew was still running projectors at The Orpheum and working at Joann's Fudge. The other guys were working in Traverse City, and Dave Pagel was back in Gaylord. It was a pretty quiet summer for us, but we did do our customary Straits Transit booze cruise in late July or early August. On bass guitar, Tom Hall, joined us for this gig. He was from Traverse City and knew Matt. Quiet and clean-cut, Tom was a good bass player. After three years, the cops had finally gotten wind of our cruises, and met us at the Straits Dock on our return, looking for minors in possession, or people who'd had too much to drink. They didn't give us a hard time.

We learned four new songs for '73, No More Mr. Nice Guy by Alice Cooper, Gimme Shelter by the Rolling Stones, Keep Playing That Rock n Roll, by Edgar Winter and Reelin In The Years by Steely Dan. It was the hardest song I ever had to learn on guitar. We debuted them at Handsome Harry's Labor Day bash at the end of the boardwalk. Carriage Tours provided a red hay wagon for a stage this year. As with

Morningstar band, beach party (September 1973)

1972, we had a Honda 5k generator for power. The party had gained steam, and Harry sold 1006 tickets for 1973, by my count. Terry Krupp, a friend of Steve's from Kalamazoo, played bass for us. He had a slight build and shoulder-length blond hair. He was excellent, and sang Gimme Shelter. He had a massive Kustom amp and dual 2x15 cabinets. He played a Gibson EBO bass. Steve used his Super Reverb amp with SRO ext cab, and I had a new Ampeg VT-40 tube amp, 60 watts, with 4x10 speakers. Steve and I debuted black Les Paul Custom guitars, mine a '67, and Steve's a '71

John Manikoff was a guest guitarist, and a fiddle player did a few tunes. Several people took some good photos, and also the Town Crier. The gig was one of our best, and a great success. It was the largest party-type gathering on the Island until surpassed by a Bayview yacht race party in the mid-2010s. The next day, Steve, Terry and I headed back to school at WMU. A day later we took three members of an all-girl rock band to the Blue Oyster Cult concert at the Kalamazoo Ice Arena, and we were right up front. The next weekend, Steve, Terry and I were joined by a drummer named Tom Smith for a gig at The Black Ram Inn, near Kalamazoo.

Throughout our lifetime, Morningstar did two original songs: a warm-up number we usually opened with, an instrumental I composed. Steve also wrote a tune in 1971 with a nice guitar hook, called Brave

New World. The 3-minute song ran about nine minutes live, with a long instrumental break in the middle.

Although no official announcement was made, Morningstar pretty much disbanded in fall of 1973. I was twenty years old, and out of rock n' roll. The years 1969-73 were memorable for me. Our major influences through our career had been Alice Cooper, The Rolling Stones, Uriah Heep, and Grand Funk Railroad. We had made three reel to reel recordings, a 1971 practice at my mom and Ray's house, Thuya, and live 1972 gigs at NCMC and the St. Ignace teen club. I returned to the Island in spring 1974, after a full school year at Western. In late August, Handsome Harry Foster was making plans for his third and final Pardner's Club party, which was ill-fated. I didn't really even want to participate this year, as our band wasn't together.

After much cajoling by Harry, I finally agreed to play, against my better judgment. The first snag came when Dan Musser of Grand Hotel, finally tired of the loud party disrupting his guests, lobbied to have it blocked from the end of the boardwalk. Of course he had lawyers. Harry and the "band" had John Croghan Sr. representing us. Mr. Musser won, and Harry was forced to move the party to an alternate location at the base of the Humbard ski hill, just off M-185 below Stonecliffe. The next bad break was, it rained all day and night, Labor Day 1974. Lastly, Harry had entrusted all the money he raised from selling tickets to a local police officer for safekeeping, but the cop left town a few days before the party, with Harry's cash. Thus Harry had no money to buy beer or pay for a dray for a stage. Somehow, we got a generator.

Singer David Nephew was still working on Mackinac, and drummer Matt Finstrom agreed to come up for the gig. On bass I recruited a friend from WMU, Rich Schroeder. He brought his friend, Jeff Bell, to play lead guitar. Our lead guitarist, Steve, was in a new band in Kalamazoo. Local friend Conrad MacDonald brought his gear and was a guest guitarist. The crowd was disgruntled to say the least, because of the rain and no free beer. A few songs into the gig, someone slipped something into our singer's drink, putting him out of commission. Next, someone cut the cord of Jeff Bell's Orange 2x12 tube amp, ending his evening. The rest of the night consisted of ad-lib jams by Conrad, Rich, Matt and me. Harry's third party ended with no one happy, and I returned to Western the next day for another year of college.

A parting thought—after our band played the Schuss Mountain job in spring of 1971, I was back in Petoskey at school when I received a call from Gene Smiltneck, head of "Bands Unlimited," a booking agency from Escanaba. They were the premier booking agency for the tip of the mitt and Upper Peninsula. Gene asked me if Morningstar would like to sign on with them. If you were a member of Bands Unlimited, you were guaranteed all the gigs you could handle for the foreseeable future. We put it to a vote, as we did with everything. Matt and I voted for joining, and the other three voted against. It was a deep disappointment for me, and told me then, the band wasn't in it for the long haul.

A postscript about my band Armageddon/Morningstar, 1969-73. I feel a band's members should all be good friends, not just mercenary musicians like Cream or The Eagles. I'm happy to say we were all good friends, and did almost everything together, especially in 1970-71. We weren't just random musicians thrown together for the sake of the music. We even had a twenty-year reunion in 1992. Don Meadows passed away on Jan 1, 1979. Steve Colbath passed away May 6, 2014.

The (Old) Village Inn: 1969-72

When the family returned from Florida in the spring of 1969 and the school year was over, it was time to find a new summer job. I went to work for my dad and godfather, Dennis Brodeur, at their bar and restaurant on the corner of Main and Hoban, named The Village Inn. It was right across from the Shepler's dock, and the State Liquor Store. My job was cooking, and morning cleanup. My fellow cooks the first year were Charlie Francis and Terry Francis (cousins).

We served primarily grilled and deep-fried food in '69 and '70; burgers, fries, onion rings, chicken, shrimp, and hot dogs. Some also came as "basket specials." We had two different pizza ovens, one large, and one small. Cooking was fun but hectic. We were popular with tourists during the daytime, and locals and workers at night. The main floor had tables and seating for about 100, and there was an elevated bar area, designed by Jack, with two server stations, which sat about fourteen. There was a walk-in cooler behind the bar, accessible from the back storeroom. Dad helped to cook when it got extremely busy in mid-summers, and Dennis helped to tend the bar. There were usually two bartenders, two or three cooks, and six or more servers per shift.

The Village Inn started out in a building at the lot adjacent to the corner, next to McNally's. In the mid-1950s it was briefly called "Pero's Pow Wow," owned by either Mort or Vern Pero. It was sold after a year or two to Bob and Arlene Chambers (no direct relation) who also added an upstairs bar. They sold the building to Jack and Dennis around 1963, who operated it for a few years before building a one-story addition on the corner lot. It had double doors onto Main Street, and a side door onto Hoban, in the raised bar area. It also featured a row of angled windows on the front of the bar, for easy, no-glare visibility. The upstairs of the old bar was converted to employee housing for servers, which Jack dubbed "The Nunnery." Jack and Dennis each had offices on site.

My dad wasn't a big music fan, but the Village Inn did hire entertainment two nights a week (later three). It was provided by the Cheboygan trio "The Zippy Zips." They were a polka band who also did current pop and traditional covers. Ray Miller, 40, played alto sax

Ray Miller and Jim Enrietti of the house band The Zippy Zips

and some tall stick device with a lever at the top connected to a string leading to a small "trash can" at the bottom. It was a primitive bass. Ray had a couple of squeeze horns connected to it. To this day, Ray Miller has been the best sax player I've ever heard. Jim Enrietti, 30, played electric accordion. He was great also. George Sanford, 20, played the Gretsch drums. I liked The Zippy Zips a lot and became good friends with the guys. Their Magnum Opus was a long version of "When the Saints Go Marching In." with a drum solo by George, when he would work his way around the bar with his drumsticks, tapping on glasses, longnecks, and liquor bottles. I taped a Zippy Zips gig one time in 1971, and still have the cassette tape.

Ray and Jim also had a pizza business in Cheboygan, and they provided all the 12" pizzas sold at The Village Inn. At the time of this writing I reconnected with Jim Enrietti through social media, after 50 years. He's still playing accordion in polka bands! One of our beverage distributors was Jack Ryerse of St. Ignace, who went on to manage the Island branch of First National Bank. In those days, Island bars got their liquor from the State Liquor Store, located conveniently for us, right across the street. One of our staff would just push a hand truck over there and load up with a few cases of various liquors. I remember that Jay Parker and Dick Hoppenrath were clerks. In those days there were two kinds of watering holes: saloons, and cocktail lounges. The old Village Inn was the former.

I got our band's drummer, Matt a job at the Village Inn and later, guitarist Steve. Don Meadows and David Nephew worked there part time, as they were primarily committed to Frank Nephew's businesses. All were cooks.

In 1972, the Village Inn converted the first floor of the old bar to a "Steak House," featuring dinners on an oval platter of a Delmonico steak, steak fries and a salad bowl, reasonably priced. Guitarist Steve Colbath became the cook in the separate, Steak House kitchen.

An anecdote from the Village Inn: Our shrimp baskets normally contained eight shrimp, with french fries. One slower evening, Dennis came walking through the kitchen and said to me, "Why don't you throw me in about six shrimp?" So, I cooked Dennis six orders of shrimp basket, not realizing he just meant a smaller single portion. So, the whole shift ended up getting free shrimp baskets that evening, courtesy of Dennis. I think Jack thoroughly enjoyed helping cook when it was busy. He either ran the grill or expedited the food going out the window to the servers. He often sang while cooking. He told me that a good summer day for the bar was when food, beer/wine and liquor each grossed $1000, and these were the days when hamburgers or a draft beer cost 50 cents.

For the 1972 season, after incessantly bugging Jack, I was promoted to bartender. My fellow bartenders for the season were "Diamond" Jim Brady, Rob "Oyster" Oesterle, and Islander Anna Joyce Putnam. Oyster and I usually had the night shift, but it varied a little. Ron Dufina had moved on, first trying a (short-lived) dry cleaning business, then hiring on at Waterways to work on the yacht dock. We carried Budweiser on draft for 50 cents, and longneck bottles of Bud, Pabst, Stroh's, Hamm's Schlitz and Blatz for 60 cents. I wasn't a very good bartender because I was not outgoing, and didn't have the personality for the job. Also, I didn't know enough to give the local steady regulars a little extra booze with their shot and mix. Nobody told me. But they all cut me some slack, and I lasted from about June 1 to the first week of October. Around Labor Day of '72 I had gotten my own room upstairs in The Nunnery, front west side overlooking Main Street. Prior to that I had spend summer '72 living at Thuya with mom, Ray and siblings, but after the previous summer on my own at Silver Birches, it was not for me.

In early October my dad, my great-uncle Tom and I took an all-day shopping trip to Petoskey. We had lunch, bought some things we wanted, and all three of us had a great time. I got some books and record albums, and was looking forward to an entertaining night in my room. We returned on a late afternoon boat and stopped in the Village Inn. Bartender Anna Joyce decided on short notice she wanted the night off. Jack asked me if I wanted to take her shift. I said no. Wrong answer. He decided then and there to let me go from my bartender job. Miraculously, I didn't get kicked out of my housing. I stayed on the Island until early January, when I left for Western Michigan U. in Kalamazoo, and my third year of college.

Jack and Dennis' partnership lasted a few more years. Dennis then purchased The Mustang Lounge from Cheboygan businessman Floyd "Butch" Walters. After the 1977 season, Jack converted The Village Inn bar to a gift shop, at the urging of his third wife, Terrie. He sold his liquor license to George Staffan, and bought his 30 ft Sea Ray yacht, the *I Aint Skairt*. I returned to work for Jack at the gift shop for about half of the 1979 season. I worked with Terrie and four girls from St. Ignace: Renata McGrath, Ann Savard, Glenda Pemble and Barb Cahill.

Silver Birches

In the spring of 1971, four of the band guys were graduating high school. I had just finished my first year at NCMC, having graduated in 1970. We were all heading back to the Island for summer jobs and looking for a place to stay. Nobody wanted to stay in employee housing, and I didn't want to stay with mom and Ray. Matt heard a guy named Rick Martin was in charge of the vacant, run down Silver Birches estate at the north end of the Island. Matt's parents knew Rick, as they had been co-workers at Mackinac College.

Rick had stayed on at Mackinac when the college closed and changed hands, buying a house with his wife, Alice, and kids Jimmy and Tricia. So, Matt approached Rick about him and two of the other guys (David and Don) becoming "caretakers" at Silver Birches for the season. Rick went along with the idea. The three of them got free rent, in exchange for their caretaker duties. When the rest of the band heard about it, we were astounded by the guys' good luck. Naturally, Steve and I wanted to get in on it, too.

The back cottage at Silver Birches had six bedrooms. It seemed perfect for us. David, and Don's best friend, Roger Miller, who also worked for Frank Nephew, was invited to make the sixth man. Rick decided he didn't really need six caretakers, so he charged the newcomers $25 per week rent.

There was also a smaller three-bedroom cottage in front near the road, and the big main lodge, where we figured we would set up our band gear for practices. The lodge had window shutters with crescent moon shaped cutouts, which we kept closed most of the time. There were snoopy tourists in those days, too, although fewer in number than today. So we got the idea to put mirrors on the inside of the crescent moons, so if tourists peeked in, they would see a pair of eyes staring back at them! Unnerving for sure, and usually chased them away.

The lodge also contained the laundry facilities for the whole place. Even though 1971 was a chilly summer, we decided to try to use the big swimming pool on the property. We kept a garden hose running into it most of the summer, but it never seemed to fill. Cracks in the bottom must have been allowing water to escape at an equal rate to the

David, Matt and Tom revisit our 1971 home about 15 years later

hose. Silver Birches either had a well or pumped water out of the lake, I don't recall which, but at any rate we didn't have to pay for City water.

Silver Birches also had a medium sized dock on the lake. It was still semi-functional, but had fallen into disrepair. It had been built by Commander E.M. Tellefson, and consisted of three large galvanized steel cylinders, filled with stone or gravel, I-beams on top, and wood planks for decking.

Silver Birches was also a natural party place. Even though we were all 17 or 18 years old, we had no trouble getting alcohol. We usually got some server girls we knew from one of the bars to "buy" for us. We pretty much always drank beer, and a little wine. A couple of the guys drank liquor, but not much. In those days, wines were limited for young people, and the choice was usually Boone's Farm. As for beer, I drank Budweiser, (because my dad did). He later changed to Miller Lite.

We weren't heavy drinkers and we didn't party excessively, because the five of us realized it wasn't compatible with trying to be a successful, fledgling rock band. Occasionally some excessive parties occurred, usually when co-workers of Don and David from

Nephew/Benser rode out there at night. Everyone was jealous that we had such a place to ourselves, of course.

None of our co-workers from the Village Inn came out to party, since they were generally older than us. Silver Birches was a daunting ride at night; four miles out of town, sketchy roads, and poor bike lights, plus alcohol or weed.

Don, David and Roger worked for the various holdings of Frank Nephew: Pickle Barrel, Orpheum Theater, Mighty Mac, and Joann's Fudge. Matt, Steve and I worked at the old Village Inn. The only one of us with a steady girlfriend was David, whose high school girlfriend, Lynne, also worked for Frank Nephew. The rest of us were "single" and dated pretty much whoever we could. Even a small, local band like ours had girls that like to hang with the band, I don't like to call them by the familiar term, groupies.

When the band had to go into town or the mainland for a gig, we'd usually hired a dray, but that got expensive after a while. One time later in the summer, we decided to save some money. We found a large cart, about the size of the recent Shepler's quarry. It only had a single axle in the center, with two pneumatic rubber car tires. We loaded it up with all our band gear. We had a lot of stuff for a young band...big amps, big drum kit. Don had some kind of leg injury, sprained ankle or something, so he offered to ride on top and balance the load. We fell for it, but in all fairness, Don wouldn't have been much use pushing the fully loaded cart. David, Matt, Steve and I pushed the cart the four miles, around the east side into town, with Don riding on top. By the time we got into town, we were a little late for the Community Hall gig, and worn out. At the bottom of Astor Street, a bunch of guys hanging around in front of the Hall spotted us and came down and pushed the cart up the hill for us. The rest of the night, and the gig, went OK.

My dad was always suspicious of our activities at Silver Birches. After all, three of us were his employees. The other three worked for Frank Nephew, including David, who was Frank's nephew. My dad owned a chartreuse colored 18 ft wooden boat with twin 45hp motors on the back. One day he rounded up Frank, and they drove the boat out to SB to pull an inspection of the place. The Mackinac grapevine being what it is, we got wind of it, and those of us who were home hid all the beer and empty cans, and cleaned up the place as best we could on short notice. There was no smell of weed in the air. Jack and Frank tied up at the dock and spent about half an hour there, looking the property over. I could tell my dad was perplexed, and maybe a little annoyed they didn't find anything. Frank just looked a little sheepish about the whole thing. Frank wished us well, and they got back in the outboard and drove back into town. The rest of our summer was incident-free.

The single band members at Silver Birches each ended up dating a girl or two over the course of the summer. I remember a local girl my age spent a lot of time at SB, and with me in particular. She was an avid follower of our band. I didn't really date a lot, being busy with the band, work, bike riding and other hobbies. I remember seeing a blonde girl, GM from Alpena, briefly. I invited her out to SB and we decided to go for an evening walk, out the back way, and down Scott's Cave Road. It seemed like a cool date, but a nighttime thunderstorm blew in from nowhere. Unbeknownst to me, G was deathly afraid of thunder and lightning. It was now fully dark, and the lightning flashes were killing our night vision. We didn't have a flashlight. Luckily, not much rain accompanied the storm. I'm not sure what made me suggest the poor idea that we cut through the woods to get back home. It could have been weed or nervousness, or G's hysterics, anyway, we cut through forest which actually was on the south side of Scott's Cave Road. Still lost, and with not much ability to see, we wandered around the woods for a half hour before finally coming out at the landfill. I was never so glad to see the dump! We made our way to British Landing Road, where a hayride just happened to be heading north along the road. We hopped on, and they gave us a ride back to Silver Birches.

Late in August, Matt and Steve invited four girls up from Traverse City, who had been their high school classmates. Roger was running projectors at The Orpheum, so the four girls hung out and partied with Matt, Steve, Don and me. As things tended to do, we became four couples for a long weekend. Everybody thoroughly enjoyed it. I ended up dating the girl I was with, a petite brunette with waist-length hair and bangs, until February '72. I used to travel down from NCMC to Traverse City on most weekends to see her.

Just after Labor Day, it was time for those of us who were college bound to leave Silver Birches. Steve went to Kalamazoo, Matt to Ann Arbor and U of M. I think David went to Henry Ford Community College in the Detroit area. I went back for my sophomore year at

NCMC, where I was fortunate to land a job as dorm Resident Assistant, "RA." I was given the third Floor. I had a single room, which turned out to be a suite, because the third Floor was not full. The first Floor RA was a big guy named Jerry Heaton, and the second Floor went to an Army vet, Ed Trudeau. I'm still friends with Ed to this day. Don Meadows stayed on at SB until late October. One night in October, Don and I got in an argument on the phone, and I fired him from the band. It was kind of a moot point, because he was joining the US Air Force soon anyway. The summer of 1971 had been a memorable, lucky experience for me, and I'll remember it for a lifetime.

Summers of 1973-74

When I came back from WMU in the spring of 1973, I needed a new job. At the time, one of the highest paying jobs in town was working for the City, sweeping streets. $5.50 an hour, I believe, so I hired on. A lot of local guys worked there that summer: Paul Wandrie, Ricky Wessel, Dale Gallagher Jr., Army Horn. Fred Hatch was still sweeping, and he did Main Street. The rest of us did side streets, Market and The Mission. My "beat" was Cadotte, from Marvin May's house to the French Outpost. Grand Hotel sweepers took over north of the OP. Mike Bradley was City foreman. He did small jobs for the City and was our go-between with the overall boss, Police Chief Otto Wandrie. I always thought Otto was a great Chief of Police, and he looked impressive riding his horse, Pepper, on duty.

If it rained hard, Otto didn't insist we sweep, so we'd go back and hang out in the city shop/barn. One time we "locked" Dale Mikey Gallagher in the old hearse, but he wasn't fazed, he just lay down and took a nap. The summer of 1973 was a sunny one, and I didn't wear a cap. The tips of my wavy brown hair actually got a few blond streaks in them! It was a good summer job, and I went back to Western right after my band's Labor Day beach party gig for Handsome Harry Foster.

When I came back from school in 1974, I spent most of the summer working for my stepdad, Ray, at Gerry and Frances Ross' house in the Annex. They were doing a lot of renovation in the large mansion, and the predominant colors were red, black and gold. We also built a new fence around the property. Gerry also decided to build a tennis court on a level area of the bluff below their house. I helped clear the land for it and was the only one of the crew who didn't catch a case of poison ivy. I got along well with the Rosses, and I always wanted to ask one of the older daughters out, but never got up the courage.

Gerry was always "simpatico" with me, as we both rode a Schwinn Paramount, top-of-the-line 10-speed. He bought one for his wife, too. One time my Paramount was in the Schwinn shop for some fine tuning, and I borrowed mechanic Paul Young's touring model Paramount to go back to work after lunch. Gerry saw it and said, "I thought your

Paramount was green?" I replied, "Oh, it is Gerry, this is my spare." Gerry's mouth fell open and you could have knocked him over with a feather. The Rosses' Paramounts eventually got stolen one winter, during a break-in.

Later in the summer of '74, Eric Yoder and I painted Pine Cottage for Little Bob Hughey. Bob had bought a bunch of five-gallon buckets of medium-brown army surplus paint. We painted all three upper floors using homemade scaffolding, made of new 2x4s hung out the windows, and wooden planks.

Ray went on to do some carpentry work for John Drudi late in the summer. John was the owner of East Bluff's "Baby Grand" then, and a good friend of Gerry Ross. Drudi also partnered with Tom Pfeiffelmann and Frank Nephew in an unsuccessful bid to buy Straits Transit. Ray was the one who joined the two halves of the Baby Grand together, with the first-floor main entrance addition. They had been separate until then. In the old days, the story was, the husband and wife each had their own wing.

Bike Races

As long as I can remember, there had been bicycle races on Mackinac. They were usually held as part of the afternoon 4th of July festivities, and went around the Island. Anyone could enter, with any kind of bike. In the 1940s and 50s it was mostly single-speeds. I recall entering one in 1973 with my new black Schwinn Paramount.

Team relay races started in 1974. They went the opposite direction from the old singles race, westward first. A bike race committee was set up. I recall Don Vanier of the Schwinn shop and Lee Finkel being on it. Bob Carr did a lot of the race photography. There were both men's and women's five-person teams. Each rider did a lap around the island, tagging the next rider in front of Marquette Park. The route avoided downtown, instead heading up Fort hill to Market Street, and down Benjamin Lane, turning right to make the lap around. "Spotters" were placed at strategic corners in town, for safety reasons. Teams were required to have a business sponsor, and they wore matching jerseys/tee shirts. There was a $25 entry fee for each team, which went toward a large trophy and medals for gold, silver and bronze. Races were held in August, late afternoon through early evening.

Neither the City nor Carriage Tours objected to cooperating with the relay race. Most racers used quality or even high-end 10-speed bikes. I put together a team the first year (1974) and my dad's Village Inn bar sponsored us. We had green tee-shirts, with a gold "VI" on the front (for the Village Inn). I recruited Tim Horn, Steve Sweet, Bruce Goodwin and Dan Johnson as our members. We had the misfortune of getting a flat tire on Steve Sweet's lap three. Steve was able to grab a bike from a summer employee to finish the race, but we had fallen behind, and didn't place that first year. Tom Pfeiffelmann's Island Bicycle Livery team took first place for the men. Team captain Paul Young rode a brown Schwinn Paramount. In the early years there were four to five men's teams and three or four women's teams.

In 1975, Ty's Restaurant sponsored our same team, and our tee-shirts were navy blue with white numbers and name. We took the first place gold. I actually wore a pair or real cycling shoes with the metal cleats on the bottom, the only rider to do so that year.

1975 relay race. Ty's Restaurant team

By 1976, my teammate Tim Horn bought a Schwinn Paramount. Our same team won again, making it two golds in a row. I think Dan Johnson had our fastest lap, on his silver Lambert/Viscount. Ty and Ling Horn showed up at the finish for the presentation of medals and trophy.

In 1977, another group of Island guys formed a new team, and they wanted to unseat us. They found a big-name sponsor, Harry Ryba, and the Island House Hotel. Their team was made up of locals Mark Chambers and Mark Bunker, along with summer employees Mike Byers, Mark Crossley, and Keith Howie, who later married Islander Sherri Hume. '77 figured to be a tight race, with Ty's trying for a 3-peat, against the new guys' top-notch team.

In my lap two, the pack of around six riders had rounded the corner onto Market St. and were about in front of the police station when an elderly tourist fellow on a single speed veered into the group, going in the same direction. My front tire hit his right pedal, and it pushed me a foot to the right. We all kept going and the elder gent did not go down. I had always used very high pressure tubular sew-up racing tires. They were the lightest weight available, but fragile. By the time I got to Arch Rock, I noticed my front tired had lost about half its pressure and fell behind the pack. Our third, fourth and fifth lap could not make up the lost time, and we finished a disappointing third place, getting the bronze medal. Island House went on to win.

A controversy developed about 1977. British Landing resident, John McCabe, wrote a letter to the *Town Crier*, stating that the bike races were dangerous, and the riders reckless. He said his mother was now afraid to take her short walks on M-185 (the paved road that circles the island). Being the only bike race member who had taken multiple English and writing classes in college, I took it upon myself to write a rebuttal and submit it to the *Crier*. McCabe replied and I replied again. Nothing detrimental to the race came from his letters, but Harry Ryba had read my rebuttals, and was impressed, so he invited my whole Ty's Restaurant team to the Island House one night for drinks on him, even though we were the competition.

1978, was again fiercely competitive. Our team had trained fairly hard and had no flat tires (as we did in 1974 and 1977!) but Island House once again took the gold. We placed second and received the silver medal.

1979 was a pivotal year. The men's race trophy was running out of room for inscriptions, and so it was determined that whoever won a third time would get to keep the 70s trophy. Our anchor man, Dan Johnson, was not on the Island in '79, so we recruited Geoff Rudolph to take his place on Lap 5. We trained harder than in any other year. Gene Arbib had bought us new numbered jerseys, in our familiar Ty's Restaurant navy blue. Misfortune struck however, and the day of the race, both Geoff and I were sick. He had the flu and I came down with mono. In spite of all our training that year, our team would have to default. Island House took their third win in a row, celebrated, and kept the large trophy.

Ironically, that year's race date had been moved one week on short notice, after a women's team had lobbied to do so. I was not aware of it until it was a *fait accompli*. Had the race been on its original date, I'm sure our team would have put up a good fight for the gold. I didn't race again after that. Three of our members, Steve Sweet, Bruce Goodwin, and Geoff Rudolph joined another team in 1980, but didn't take first. I believe it was The Pilot House. Teammate, Tim Horn, became a race official in 1980. Strong teams who had wins in the early 80s were May's Fudge and Island Bicycle Livery. The relay race continued until the mid-1980s, then The City and Carriage Tours decided it was no longer in their interest to cooperate. The City cited insurance concerns, and Tours maintained it was just too disruptive to their carriage operation.

Some members of women's teams I recall through the six years I was involved were Valerie Porter, Lynn Pope, Suzanne Plunkett, Moira Croghan, Midge Bodwin, Laurie Chase and Lydia Goodwin.

One year, I think it was 1978, our Ty's Restaurant team did a day of training on the mainland. The five of us caught the 8:00am boat to St. Ignace. Mark Chambers asked if he could join us. Our ride was to Hessel and back, with each of us taking turns leading, and our route was Mackinac Trail to M-134. We spent an hour in Hessel for lunch, and on the way back stopped at Grant and Jane Simmons' house for a few minutes. Daughter Allison was on the porch getting some sun. We managed to make the 3:00pm boat back to the Island, so we made excellent time on our 50-mile ride.

One year in the mid 1980s, there was even a race for individuals that went up Mission Hill and through the center of the Island. I think it only was held one year.

In around 2010, former Island House team member, Mark Crossley, got in touch with me and asked if I'd like to caretake the old Mens Relay Race trophy. He had it for many years, and no longer had room for it. We met on the Arnold Dock one summer day, reminisced for a few minutes, and his daughter took a photo of us with the trophy. I still have it to this day.

Working For "The General"

I was looking for a new job in the fall of 1974. Contractor Jim Francis was hiring. He ran a crew of two or three carpenters and four or five painters. He took care of over a dozen cottages on both bluffs and in the Annex, and also did work downtown. He had recently built the new French Outpost, and Selma's Fudge. When my great Uncle, Chuck Dufina, retired and passed away, Jim had bought his equipment. I hired on the paint crew.

The foreman was Joseph "Snapper" Bazinau. He was planning to retire from painting in 1975, but I could learn from him until then. His crew and others had nicknamed Jim The General, because he was the leader. Most years Jim's painters didn't work winters, so we had about another six weeks until layoff. One of our primary jobs in fall was to close cottages, sometimes in the evening after regular painting hours. Jim would drain the water pipes and the paint crew shuttered the house. In the spring we reversed the process. Everybody used shutters back then, and they were either stored in a barn or under the cottage. Large houses like Dziabis' or Cochran-Hemmingway's (now Levy) took most of the day. I remember counting 144 shutters for Hemmingway's. The old wooden ones there were extremely heavy, but Jimmy Francis, Jim's son, was in the process of building new ones out of 3/8" plywood and painting them white.

The crew was transitioning, and by spring it was made up of Loren Horn, Armin Porter, Bobby St. Onge Jr, and me. Jimmy Francis worked part time, as he also ran a 28-foot charter boat, the *Vera Mae*. Carpenters were Jim Sr.'s brother, Vin Francis, and Wilson LaPine. Over the winter ,Jim lined up a series of houses or businesses for the crew to paint the following season when the weather permitted, May through October, I worked for Jim through the end of 1978, and again late in 1979. I returned to work for him one more time, when Jimmy and I painted his house. He was retired by then. I have lots of stories and memories from my career with The General.

Jim had a gruff personality, and often chewed out the crew. His targets were Jimmy, me and Bobby, in that order. We could never tell if we had actually done something wrong, or if it was just his way. He was intimidating, and you could always tell when he was coming to a jobsite. He was preceded by his dogs Rufus and Shadow.

In the springtime, there were often bats behind the shutters, and some of the guys were nervous about them. It didn't help if you had smoked some weed, which some of us did to relieve the boredom of shutter duty. One job, we split up, Bobby and I going to paint a set of porch steps at Charlotte Sweeney's, and Loren and Armin going to Puttkammer's to start taking down the shutters. My thinking was, by the time we finished the steps and joined them, they would have half the house done! But no. We got there and I asked the guys, "Why don't you have any shutters down?" Armin's reply, "Loren thought he heard a bat."

One year we had the contract for painting the east side of Chippewa Hotel for Nate Shayne. It was wood siding then. It wasn't practical to use Jim's big 40-foot or 48-foot wooden ladders, so we used a device called "the swinging stage." It consisted of a 16-foot aluminum plank, raised and lowered by a pulley system and heavy ropes. It was practical if the roof was flat. The swinging stage was secured to the roof by heavy iron hooks, about two feet in length that looked like giant fishhooks. These were called "sky hooks." Guys good with heights manned the plank and painted, and two other guys worked the ropes, raising and lowering the device.

As we were all on the ground taking our 3pm coffee break, black clouds were gathering overhead. Jim showed up, and Jimmy and I, always on the lookout for rain, told him, "It's gonna rain." Jim told Loren, "I don't know, Loren, you can try it if you want." Loren never liked to miss any work, so up the stage they went. A few minutes later, the skies opened up. Bobby and Armin put away tools and Jim and I lowered the swinging stage, with Loren and Jimmy on it. I had never learned sailor's knots, I just tied anything that would hold. Jim was working the left pulley rope, (which I had tied) and getting drenched by rain coming off the roof. "Who the hell tied this knot??" he yelled.

Some of the cottagers we painted for also hired seasonal domestic help and housed them. One time while painting a house I noticed their male domestic helper using a roller (instead of a brush) to paint porch furniture. I never understood that choice. When closing a cottage at the end of the season, the owners usually told us to help ourselves to whatever was left in the fridge. My favorite place for this was Porritt's cottage, Mike Hegarty's place on the East Bluff.

Big Mike always left us great stuff, one year even the better part of a standing rib roast. Another time, Jim got there ahead of us, and all the crew ended up with was tonic water. So it became a game for us, to try to get to a cottage ahead of Jim, for the leftover grub. Ted Sweeney's also had good things left in the fridge.

One day early in my career, we were painting Hazlett's (later owned by Musser). Snapper was still with us. It was so hot, 90F, that we all decided to pack up and leave, except for Loren, who continued, painting the blue porch ceiling.

Occasionally the crew took on jobs other than painting or shuttering. One year Jim was hired to pour a large concrete sidewalk and support wall at Louis Bodek's. It was the only one-story house in the Annex, a bit newer than the rest, and surrounded by woods, so that you could not see it from roads on either side of the property. Our five-man paint crew did the mixing, hauling by wheelbarrow, and troweling of all the fresh concrete. The house is currently owned by the Michael Straus family.

Jim sometimes hired guys for short time periods when they were between jobs, usually older guys from the Island he knew. George "Hong Kong" Bazinau was one of them. The crew was painting Straus' big yellow house, and I was painting several French windows on a rounded area on the first floor front. Hong Kong showed up, spotted me and said, "Let's set something up." I was already set on an extension ladder and didn't really need to change anything. I humored him anyway. His idea was to set up two ladders and a 16-foot aluminum plank. I told him it wasn't going to work. The plank was way too long for the rounded section of windows. The ends would be too far out from the building. While we were standing there on the lawn trying to figure out what to do, here came Jim. Oh no. He greeted us with "What in the hell are you guys doing?"

Another anecdote: one time we were painting for a family whose eldest son had the task of mowing the lawn. He disliked the job, and reasoned if he didn't put oil in the mower and it broke down, he wouldn't have to cut grass anymore. His dad figured it out and gave him a severe reprimand.

For the summer of 1978, Johnny Fisher joined Jim's paint crew. I remember him being with us when we painted Roy's on the West Bluff.

One of the more unusual jobs we got while I was working for Jim Francis, was at the "Power House" on M-185, about a mile out of town on the east side. An alternate, seldom used name was The Water Works. In the old days it provided water/power for some of the Island. A large pipe ran up the hill behind it to a reservoir at Fort Holmes. It was a sprawling one-story structure, and contained among other things a "chlorine room" and large Nordberg diesel generator. At one time it even had a very tall chimney. There was a house on the south side of the property for the plant attendant. The last family in residence was Bud and Joan Bradley and kids. Jim sent Bobby St. Onge and me out there to paint several large silver fuel tanks on the north end of the property. The whole complex was eventually torn down and replaced. The Bradleys' son, Michael, stayed on at the house for a few decades.

Sometimes when the crew was working in The Annex in summer, it would get hot, of course. Some days for 3 o'clock break the guys would take up a collection for beer, and one would ride to town to pick it up. I was usually elected, since I had a ten-speed with a big basket. One summer's day I went to Harry Ryba's General Store to pick up the 12-pack. It was located where Mary's Bistro is now, and did a good summer business. While I was standing in line in the busy store waiting to check out the 12-pack, who do I see in line ahead of me but Jim Francis himself! What were the odds of that? I figured I'd just be low-key and he wouldn't notice me. But no. All of a sudden he turned around, surprised. "What in the hell are you doing here?" he asked. Not much really came of it, though. I bought the beer and returned to the job.

Amusing things often happened while we were painting somebody's house. One was at Harry Ryba's house, now owned by Nancy Porter. As Ryba's Fudge box color was pink, they had several pink rooms, either wallpapered or painted. Many rooms had white carpet also. Two of us were inside one day, to access the second floor front porch through the front bedroom. I happened to overhear Harry on the upstairs phone agitated and clearly arguing with someone. When he hung up, his wife Ethel asked him, "Who was that?" Harry replied, "Some a**hole who claimed he cast the deciding vote for me to get the liquor license at the Island House. He wanted a free hotel room for the weekend. I told him to go get f*cked!" Of course we painters had to control our snickering.

Mackinac Living, 1973-1980

I looked for a new place to stay in the spring of 1973 when I returned home from Western Michigan University. My grandfather, Phil Dufina, had had a small cabin at British Landing, which was now sitting vacant. It sat on a hill on British Landing Road, on a small triangular wooded lot. It was basically your old proverbial "tarpaper shack" but the structure was good. My Uncle Dennis now owned it, and I asked him if I could fix it up, with help from my stepdad, Ray. The house had a large living room in front, a kitchen, a small bathroom, and a nook in the back corner big enough for a bed and table.

Ray had new electrical service put in, and did the wiring and plumbing. My friends and I painted the rooms. It remained with black paper on the exterior, we never got around to siding it. Having a quiet location out of town, it became a popular place for my friends to hang out. Several of my classmates from WMU also visited in summers. It was not winterized, but I did spend four summers there, 1973-76.

Uncle Dennis even offered to sell it to me for $5,000. I didn't have the money at the time, but in hindsight I wish I would have taken him up on his offer. In late summer 1976, the cabin was broken into, and a few things stolen. I decided it would be my last season there. My studies at Western were now completed and I was still working for Jim Francis, so I got an apartment in town for the winter. I rented unit 3 of the Breakwater Apartments from Tom Pfeiffelmann. The four-unit building of cinder block construction had been built by John Bloswick, and was known as the "cell block." Other residents were Bobby St. Onge Jr., Dan Johnson, and Dave Pulve of Doud's Market.

A new opportunity arose in spring 1977. First, a little backstory. The Preston house, just east of Ste Anne's Church, had been vacant for years, since the passing of Marjorie Preston in the 1960s. I remember her from when I was a child in the neighborhood. Colonel William P. Preston (1845-1916) had been a prominent figure in Mackinac County and on the Island. He was also mayor of Mackinac Island for 17 years. He had been a Civil War veteran and also stationed at Fort Mackinac.

Painting of my grandfather Phil's cabin at British Landing. The dog's name is Patches.

The Preston family also operated a fountain-style shop downtown, in the location now known as The Thunderbird.

Their large, L-shaped house with Greek style columns now had broken windows, shutters askew, and tall weeds covering the yard. It looked like your proverbial "haunted house." It was purchased in 1976 by Vernon and Joyce Haan of Lake Bluff, Illinois. Vern hired my stepdad, Ray Summerfield, to do the renovation on the house, one of the oldest buildings in the Mission.

I did a lot of the painting for Ray, and the Haans, in the three-bedroom main area of the house, which included the front entrance. What struck me were the low ceilings, only about seven feet. A new efficiency apartment was constructed upstairs, on the northeast corner of the building. It had a nice kitchen, a new bathroom, and a large living room with a sofa-sleeper. I rented it from Vern and Joyce from the time it was ready until late 1979.

The entrance was at the back of the house, with an interior stairway. The "west wing" of the building was put on hold for the time being. Vern Haan later remodeled it himself. The Haan family could not spend a lot of time on the Island in the first few years, so they rented out the main house, weekly or monthly.

I was often in charge of collecting the rent, especially in 1977. One early tenant was Tom Lewand and family, who later purchased an East Bluff cottage. I enjoyed the three years I lived at Haans' and became friends with their sons David and Nicholas. Their middle son, Bradley, didn't visit the Island often. I left my apartment there for good in November 1979, to return to Florida for winter work. Haan later put an addition on the north side of the house. Vern did much of the work himself and used a lot of old logs in the construction process.

The year 1979, of course, had been the summer of *Somewhere In Time*, the second major movie filmed on the Island. As did many Islanders, I watched the filming, and managed to get Jane Seymour's autograph. My mom and Ray hosted some parties at their house for the cast and crew. Christopher Reeve was tall, 6'4", and still in good shape from playing Superman. One of his early trips to the Island was on my dad's yacht *I A'int Skairt* in order to avoid ferries packed with tourists. The cast and crew all rode lime green Schwinn 3-speed Collegiates, with a 21" frame. Reeve's was much too small for his build.

When I returned from Florida work in spring 1980, I learned that my stepdad, Ray, was doing work on the East Bluff for the new owners of Sunrise Cottage, Jerry and Sandy Brady. They had bought it in 1979. The Bradys had a daughter, Liz, a few years younger than me. Ray put together a paint crew for Sunrise, made up of me and several friends. The cottage was white in those days. The Bradys also rented me an apartment over their barn. They were to go through a divorce soon, and were not on the Island a lot. Sandy Brady gave me a room inside the main house to stay in when they were away. Mrs. Brady was also a good friend of my mom. Sunrise was a wonderful place to stay, and the highlight was when I threw an outdoor barbecue party near the end of the summer for about 20 of my friends, guys and girls alike. Early the next morning I received a call from Sandy Brady, saying she would be up by early afternoon! My buddy, Dan Johnson, came up to Sunrise and we made sure everything was cleaned up from the night before.

I had spent much of the season of 1980 working for Matt Porter's new "Straits Construction" company. It was his first business, starting up the year before. Some of my co-workers were Herb St. Onge, Mark

Bunker, Mike McCord and Robin Hayes. When Sandy closed Sunrise in mid October I stayed with my good friend, Armin Porter, for a few weeks at his Rowe 2 apartment. He had recently gone through a divorce from his first wife. The first of November I was off to Florida again, for my second year of winter work there.

Dabbling in Firearms

Firearms were and are popular with quite a few people on Mackinac. It was probably more the case in the old days. I didn't come from a family of hunters. Neither my dad nor step-dad hunted. None of my close uncles or cousins did so either. But several of my friends hunted, and were also into target practice shooting: Tim Horn, Chris Bloswick, and Jimmy Francis. They each owned several rifles and pistols. They hunted on Bois Blanc Island and shot for target practice at the Mackinac Island landfill. Up until the early 1980s, you could also target practice on Round Island in the winter, if the channel was free enough from ice to allow travel by a small outboard boat. So naturally I started to want to tag along. I never did own a rifle, but I did use one each from Tim and Chris.

I did however manage to come by two pistols in the mid 1970s. The first was a Walther P-38, 9mm, from WWII. It had a clip holding 8 bullets. It had all matching parts serial numbers, which was valuable. It impressed Tom Pfeiffelmann who was a collector of German artifacts. My dad had come by it in North Africa, while with the Merchant Marine, right after the war. He gave it to me, under the condition that I first learn to take it apart and reassemble it in one minute, which I did. The modern 9mm shells are copper-jacketed and hard on an old WWII pistol. So, Tom recommended I order lead ammo for my target practice.

Also in the mid 70s, I decided to order a "Buntline Special" from Ace Hardware in St. Ignace, (Harrington & Richardson model 676). It was the pistol made famous by Wyatt Earp, and I had avidly watched Hugh O'Brien portray Earp on TV when I was a kid. It was a revolver with a 12-inch barrel. Mine had interchangeable cylinders: 22 long rifle and 22 Magnum. The magnum shells were longer and the manual said it had a range of 1.75 miles. I used both pistols for target practice with my friends, but they didn't like the 22 Magnum, as when fired it made a very loud "crack" sound. I ended up selling the Buntline in Florida, 1981. The Walther ended up going back to Jack.

The Day the Boats Almost Quit

It was the third Saturday in September 1978, and a strong northeast wind had been blowing for two days, a real "Nor'easter." It was a bit early in the fall for this, but I'm thinking a powerful low-pressure system had parked itself over Ontario. I was living at Haan's that year, and had the day off from Jim Francis' paint crew. I had always loved riding the boats in rough seas, so I stopped by Michael Hart's place at French Lane to see if he wanted to go take a ride. He had the day off as well. I said to him, "Let's go for a ride with Tut."

He was up for it so we headed down to the Arnold Dock, for about a 10:30 boat. This was the first year Straits Transit boats were landing at the Arnold Dock. *Straits of Mackinac II* was using the old Mackinaw City ramp on the west side, as was the practice then. When we got there and said hello to Tut, he said "You guys have got to be crazy to want to ride in this weather." Shepler's had shut down for the day. There was no Star Line yet.

It was cloudy, and although I did not check the morning TV weather, I would guess sustained winds were over 30mph with gusts of 50. No boats were landing at the big State Dock in Mackinaw City that day. Arnold Line had not yet started landing vessels at the old Straits Dock. We were at the Straits Dock in Mackinaw.

I heard much later that Paul Allers had the day off. Mike and I rode on the upper deck. We were the only ones Tut allowed up there that day. The ride to Mac City was fairly uneventful, as the seas were directly at our stern. No passengers were aboard. There was just a slight pitch and roll once we "surfed" through the channel. In normal conditions, Tut ran the *Straits II* fairly fast, and she had the most speed of any displacement hull vessel in the Straits. He could make a run in 29 minutes.

We tied up in Mackinaw, were rolling a bit at the pier, and I noticed green water surging halfway up the two stern windows. We boarded about forty tourists, and the crew gave them instructions not to go up top. Next, two carts were brought aboard, one of luggage and one of milk and other dairy products.

Mike and I headed upstairs again, and Tut got *Straits II* underway. We headed toward the Island, pitching into a head sea, at about 12 knots. There was a lot of spray coming over the bow, and the wind was blowing it the full length of the vessel, so we hunkered down behind the pilot house. The wind was also making a loud whistling sound and blew through the *Strait*'s mesh railing on the upper deck. In those days, many of the ship's life jackets were stored underneath the roof, held in place by a network of bungee cords. The strong northeast wind was blowing these loose, and they were flying down the deck toward the stern. So, Mike and I kept dashing out from behind the pilot house and grabbing loose life jackets. It was hard to maintain our balance, weaving, staggering along the upper deck.

We stowed the lifejackets as well as we could, in the space behind the pilot house. *Straits of Mackinac II* is lighter than the Arnold vessels, and she tended to rise up on the whitecapped rollers, and then drop down in a trough, rather than plowing through. This really became evident the closer we go to the Island. The stretch from old Round Island Lighthouse to the harbor entrance was the roughest, especially the channel.

I couldn't accurately measure wave height, but I'm sure they were 12 feet. Tut checked down as we made our way through the cut. I chanced peeking out from behind the pilot hose as we dropped into the trough of a swell, and all I saw was a wall of water ahead of us. It seemed like a virtual tsunami. *Straits II* rarely takes a wave over her sharply pointed bow, but we may have that day. There was so much spray and howling wind gusts, that you would have gotten instantly soaked, without the shelter of the wheelhouse.

We entered into the harbor near the East Breakwall and made a tight turn. We pulled into the west side of the Arnold Dock, and the deckhands secured the lines. As Mike and I descended the stairs, I noticed it was noticeably warmer, humid, and smelled of seasickness. Most of the forty aboard were tourists. The cartload of milk and dairy had broken free, and there was spilled milk and containers all over the deck. The *Straits II*'s starboard side sliding door opened, the ramp was put ashore, and the shaken passengers unsteadily made their way to the dock.

We said goodbye to Tut, having enjoyed the nautical roller-coaster to the max. I think *Straits of Mackinac II* made five round trips that day. I heard much later that Ray Wilkins had made a trip or two from

St. Ignace with *Huron*, but I didn't see them. Imagine these days, in the 21st Century, if there were only a half dozen boat trips to Mackinac, on a day in tourist season.

Travel 1978-79

In mid-December 1978, I flew to Great Britain to meet up with my girlfriend at the time, who was enrolled at a University in London. Christmas break was coming up for her. I first spent two weeks exploring London as she showed me around. After that we took a train to Dover on the English Channel coast. From there, it was a passenger and car ferry across to Calais, France. We traveled by train to Paris, and spent a day touring, seeing the Eifel Tower and Notre Dame Cathedral. We took the stairs to the top of Notre Dame.

We had a fourteen-day Eurail pass, and decided to make the most of it. Some nights we slept on the train, and others we got a hotel or B&B room. From Paris we went to Munich, Germany and toured the Olympic grounds, stadium and 955-foot-tall Olympic Tower. Our train ride took us along a scenic stretch of the Rhine River.

Next was to Switzerland, as it was my girlfriend's favorite country. We stayed in Lucerne, at the small Hotel Gambrinus. Our room was on the fifth floor, and had an old balcony. The only other things on this floor were the laundry facilities for the hotel.

There was a nice long footbridge across Lake Lucerne, which I enjoyed walking. We also climbed a small mountain road to tour Chateau Gutsch; "The Gutsch," a medium-sized, pretty, older luxury hotel. From Switzerland it was on to Salzburg, Austria, which I liked. It was the setting for the popular movie *The Sound of Music*.

In every country after France, there was a bit of snow on the ground. We also wanted to see Berlin, so we took a train ride north. At that time the Eurail train pass was basically unlimited, except for the duration you had purchased, in our case, two weeks. In 1979, Germany was still divided into West and East. We toured West Berlin for a while, visiting shops and seeing an old church that had not been restored much from World War II. We both decided to visit East Berlin, into the communist part of the city, and crossed over at "Checkpoint Charlie." A few American troops were on duty there, and a soldier told us, "Don't try to bring any East German Marks (money) or coins back with you, or I guarantee you will be detained anywhere from two hours to two weeks." We spent a couple hours in the East and spent the paper money we had converted.

On the way back, near the checkpoint, we encountered a dozen or so East Germans who waited there for departing tourists to toss them unspent East German coins, which my girlfriend did. The GI told us this was a common practice for poor folk from the East. From Berlin we had a long train ride to Spain, our last country.

It was warmer in Barcelona, the low 60s. While there, we took a ferry on a harbor cruise. The US Navy helicopter carrier *Iwo Jima* was in at a nearby dock. After that we walked around the city. We climbed a hill to the fortress known as Montjuïc. There were armed soldiers there as well as around town, where most stood at attention with rifles at government buildings. It was a bit surprising to me. Our visit was during the early years of King Juan Carlos I's reign.

When we left Barcelona, we took take the train back to the English Channel, and a ferry across to the UK. I spent one more week there before saying goodbye and catching my British Airways flight back to the US. I was happy to have seen Europe once in my lifetime.

A few weeks after I was back on Mackinac, some friends and I were contacted by someone from Straits Area Tourism, in mid-winter 1979. I forget exactly how this came about, or why we were chosen, but Tim Horn, Michael Hart, Jim Francis Jr., and I were asked to represent Straits Area tourism at a winter Tourism and Sports Show, held at the mammoth McCormick Place in Chicago. As none of us had a suitable car to take, we borrowed one from our friend, Pat Granning. There were other people from St. Ignace and Mackinaw City who also worked the show, with rotating crews.

I think the Show was around 10 days, and we staffed a booth for three days, then spend two nights in a nearby hotel. We greeted the interested folks and also handed out brochures, and pitched the merits of vacationing in our area. It was two of us working at a time, and the other two could wander around and sightsee all the sporting goods and tourism stuff in McCormick. There was an option of wearing colonial garb costumes. I tried one for one day. One night we met up with David Albee of the East Bluff, who lived in Chicago. He showed us his apartment in an old multi-story building, then took us up to the roof, where David lit off fireworks for a few minutes. Police cars were soon prowling around down below, but we were gone by then.

It was a fun trip with friends in the middle of an Island winter, but it was marred by car trouble on the return drive. We had to stop at a Grand Rapids garage for parts.

Trying My Hand at Cooking

After the 1979 season was over, I decided to go south for the winter. I needed to make some travel money, so worked at Grand Hotel for two weeks, painting for Richie Bos, the architect, who was running the paint crew for his dad, Louie. It was a cold November, we could barely heat the hotel rooms adequately for painting.

In late November I left the Island and met my girlfriend at the time in suburban Cleveland. She had been a hostess at the Island House the previous summer. Our plan was to go to Clearwater, Florida together. We rented an efficiency unit at a mom and pop motel on Clearwater Beach to be close to the water. They still had those then, not having been torn down yet to make room for high-rises.

Clearwater Beach is about the same size as Mackinac Island, except long and narrow. It's connected to Clearwater by Memorial Causeway. I applied for a maintenance job at Holiday Inn Gulfview, on the south end of the Beach, about a block from where we lived. The person doing the hiring told me what they really needed were cooks for the upcoming tourist season, so I agreed to give it a try. Holiday Inn Gulfview was a large L-shaped hotel, with two wings of about seven stories each, and 290 rooms.

It was a member of a group of about ten hotels called "Johnson Properties," based in Atlanta. There was a central lobby, with a coffee house off to one side, a cocktail lounge with nightly bands, and a dining room overlooking the Gulf with seating for 156. The kitchen was big, too, with a walk-in cooler and freezer, dry storage rooms, a salad and dessert pantry, dishwasher area and chef's office. The cook's "line" had separate broiler and sauté stations, with stoves, fryers and reach-in refrigerators. Behind it was the prep, buffet and banquet area with tables and two convection ovens. I was given a night line cook job, but was expected to learn the other stations over the course of the winter.

Some of our popular entrees included fresh fish, prime rib, lobster tails, New York Strip, veal oscar, and linguini romanesca. An older lady named Virginia ran our salad and dessert department. We had about a dozen desserts, but the most popular was the "Potion of Love

Pie." It had the popularity of Grand Hotel's Pecan Ball. Popular salads were the Fathom Salad, which contained different types of fish, and the Antipasto Salad, which had over ten ingredients. Both were served on a large oval platter.

Our dining room was called Fanny's, and the menus had fancy green leatherette embossed covers. The wait staff was very professional, and all female my first year. They dressed well, (there was no uniform) and on Saturday nights, for our "Hawaiian Luau," They wore a leotard type top, long floral skirts, with Hawaiian leis and a fresh flower in their hair. We also had a hostess and several busboys. Our chef the first year was Richard Drouin, whom I got along well with.

Our kitchen had both full electric and gas cooking facilities. One time on Thanksgiving 1979, a ship hit the gas line under Memorial Causeway Bridge and knocked it out. Several restaurants on the beach had to close, and we got a lot of their holiday customers. I did a double shift and we served over 1000 dinners between 11:00am and 10:00pm.

The manager was Bob Jones, a tall blond, middle-aged guy with glasses. He was also a corporate VP with Johnson Properties. In late winter, we received an award for being one of the Top-10 Holiday Inns in the world. Management closed the dining room for a night and threw us a party. We all received black and gold pins, with "JP" on them. In April of 1980 I gave notice and headed back to Mackinac for the summer season.

I had liked my return to Clearwater and Clearwater Beach, and decided to go back after Mackinac Island's season was over. I easily got rehired at Holiday Inn Gulfview, and this year there was a new Executive Chef, George Pastor, a wide-shouldered man with brown hair and a moustache. George and I hit it off right away. The sous chef quit a few days after I arrived, and George gave me the number two spot. Hotel management didn't want to give me the title "sous chef" since I had not gone to a Culinary Institute. My name tag simply read, "Tom Chambers, Night Supervisor." Instead of line cook I was put in charge of the nightly buffet, which was a lucrative post. Virginia from the pantry was still there, and she handled buffet salads and desserts. Occasionally I also did the breakfast buffet. I considered it a nice buffet setup, with six enclosed steam tables in a counter top for entrees and veggies. On Saturday nights and holidays, there was a small table out front, where a cook would carve some type of meat, usually steamship

Cooking at Holiday Inn Gulfview 1980

round beef or turkey. There was still the Hawaiian Luau on Saturday nights, and a seafood buffet on Friday, which featured a full grouper fish all decorated up, on a bed of lettuce and trimmings.

In December 1980 I decided to buy my first car. After an unsuccessful bid to buy a 1974 silver-blue Ford Thunderbird with a moon roof, from a large dealer in downtown Clearwater, I visited a small used car dealer on the north side of town near Dunedin. My family was wintering in Dunedin, and my step-dad Ray went to the dealer with me, as he had experience from a long ago job as a used car dealer. I picked out a 1973 Lincoln Continental Mark IV, in dark Gold Moondust, with a white vinyl top. It had the Lincoln 460 four-barrel engine.

I paid $2500 for it. As I didn't have any Florida credit, a good friend of mine, a Holiday Inn bartender named Libbye, co-signed for me. When I drove it to work, Chef George wanted to check it out. When I opened the hood he exclaimed, "Look at the size of that radiator!" I had a side job at Holiday Inn Gulfview that year, as doorman/ID checker at their bar. The job required a sports coat and tie, and some of the server ladies commented how well I "cleaned up." While doing this I met one of our nightly rock bands, Fire Tower and ended up purchasing a favorite guitar of mine, a 1966 Mosrite from their rhythm guitarist, Steve. He wanted a Fender Stratocaster instead. Unfortunately, his new Strat got stolen out of the band's van a few weeks later. I had stayed on at Holiday Inn through summer, since I had a good salary, and was paying off a car. 1981 would be the only summer I ever missed on Mackinac Island.

In mid-July, Chef George got called in one day on his day off. This was unusual. He had a meeting with management, and a short while later I saw him grimly packing up his cookbooks and cleaning out his office. It turned out, the new chef and sous chef who had been hired for the new Holiday Inn a mile north on Clearwater Beach, were taking over. The new hotel was not ready, and they were under contract, Johnson Properties having promised them a job. So, George was fired, and I was demoted back to line cook. I said to myself, the hell with that, and gave my two-week notice. As I was not ready to head back to Mackinac yet, I spent the next six weeks painting a small house for one of our busboys' parents. My two years on Clearwater Beach I also cooked part time in a small restaurant right on the main drag, Mandalay Ave, called "The Ship's Pub." It was owned by Gary Stobbe, and reminded me a lot of Mackinac's Chuck Wagon. It was the same size and layout, but it also served beer and wine. The specialty was beer-battered grouper fingers, deep-fried. Libbye often stopped in for a beer when she was off-duty.

My second year on the Beach I rented a one-bedroom bungalow on East Shore Drive, about a mile from work. It had two screened porches and a backyard and carport. It gave me a view of the harbor, looking toward Clearwater proper, and Island Estates. It cost me $75 a week. During tourist season, I often rode my Schwinn Paramount to work, as it was faster in the traffic jams. That little bungalow complex was torn down a year after I left, to build a hi-rise.

I headed north in my Lincoln on Labor Day weekend, 1981, for a short fall season. I arrived in Mackinaw City around midnight. Bob Benser was just heading across in his small yacht, but it was full, so I stayed overnight in Mackinaw with my buddy, Ed Trudeau. I worked two months on the Island, and then packed up for Florida again. Brothers Doug and Gary McDonnell hitched a ride down with me.

My third winter in Florida (1981-82) I got a job as dinner cook at a large family restaurant in Dunedin, on the mainland just north of Clearwater. It was called "Some Place Else." It had three dining rooms, (one for banquets) and a separate bar area. We were low-priced and catered mainly to families. We had a parent restaurant, the Four-Star "Bon Apetit" on the Dunedin waterfront. We specialized in half-chicken dishes, served with a variety of toppings and sauces, but had a full menu. We also served lunch and breakfast. We had a maître d', a Chinese fellow named Steve Meng. The chef, Gary Jacobsen, was a year older than I was, and his wife, Terri, was a server. Gary seemed under a lot of stress, and it appeared he drank a lot. He was a difficult chef to work for.

In the first week I was there, we lost our longtime salad person, a girl named Leeds. We would also soon lose our hea cook, John Pryor. So I rose up, and soon was head dinner cook. We had to make a variety of sauces almost daily, and just before dinner hour, Gary would go through tasting the sauces. I don't know how many times I heard "This is no good, throw it out." just a few minutes before customers started arriving. So it was frustrating, and became stressful for me as well.

After a night's work, some of my co-workers and I went to a neighborhood bar called "The Regal Beagle." Yes, same as the TV show bar. Chef Gary also recruited me to become the Sunday breakfast cook. He wouldn't let us cook eggs on the grill, but insisted we use 8-inch frying pans for each order, flipping the eggs for over-easy, etc. I broke a lot of yolks at first, and it was maddening, but I soon got the hang of flipping. The cooks did most of the ordering for Some Place Else, and I was put in charge of produce orders. You would just get on the phone and call in what you needed. There was no salesman who came around daily, as on Mackinac. I lasted the whole winter there, believe it or not, and in spring management informed us the restaurant was being sold. Chef Gary got us all together and offered us three choices: staying on for the new owners, transferring to Bon Appétit, or finding a new job. It seemed like a weight was suddenly lifted from his shoulders. As spring was in the air, I chose to head north to Mackinac again.

I went back to painting, in a new business run by Loren Horn, for a year and a half, but also cooked lunch and did prep part time for Ty's Restaurant in 1982 - 83. Tim Horn was the breakfast cook, and his brother, Eddie, worked lunch. I met a new girlfriend in early summer 1982. She was best friends with Nancy Nephew Porter. We got a house together for several years and were in Nancy and Armin's wedding.

Iroquois on the Beach:
Working For the Macs, 1983-1993

After I finished working for Loren Horn's paint service, following Yacht Races 1983, I painted my mom's front fence at Thuya (The Red House) for something to do. Yes, it had a fence back then. One day, Mrs. Margaret McIntire was walking by, and commented, "Tommy, we should have you paint for the Iroquois." Within a few weeks I had accepted her offer.

The McIntires, Sam and Margaret, and my parents go way back. My mom and Margaret Davey worked together at Herpolsheimers Clothing Store (now Pancake House) in the late 1940s. Mom was in her late teens, Margaret about three years older. Margaret had a younger sister, Lettie, who was also good friends with my mom. Lettie and her husband, Joe Abbey, opened a clothing store in the Murray Hotel building in the early 1950s, called The Scotch House. It featured high-end wool plaid clothing. Sadly, Lettie passed away at a young age. Joe Abbey continued to visit the Island and stay at The Iroquois until around 2010. The Abbeys had a son, Jimmy, who was my age, and worked as chef at the Iroquois for a year or two while I was working there.

Sam M. McIntire had come to the Island as a State Police officer on Governor G. Mennen Williams's bodyguard. He met Margaret Davey, they fell in love and got married. In the early 1950s they bought Iroquois Hotel from Alicia Poole, then had four kids, Mary K, Becki, Aaron (who was my age), and Marti. Sam was a little taller than I was, in good physical shape, and had big hands, as I recall from shaking hands with him often. He originally hailed from Missouri, and never lost his Missouri twang. He became mayor but chose not to stay in local government/politics for long.

His old boss, "Soapy" Williams, and his wife Nancy, bought a house on the West Bluff when his time as governor was over. He always liked to wear a polka-dot and navy bowtie. Sam and my dad had met in the mid 1950s and became fast friends. Sam was a few years older than Jack. I might mention, Sam was also an exec in Star

Line Ferry, having gotten in on the ground floor in 1979. Sam was also an avid golfer, and occasionally he and my dad would play. One year, for the Sysco food service golf outing to be held in Petoskey for all their Mackinac customers, Mr. Mac invited me to go. I thought I'd be accompanying him. "No" he said, "You're representing the Iroquois this year, I won't be going." Petoskey BayView had a beautiful golf course, and my team was Ron Steensma, Leo Kryzminski, and a young guy named Steve I didn't know, but could hit the ball a mile. We didn't end up winning anything.

The Macs were great people to work for...fair, smart, and good to their employees, yet also demanding. My job was small hotel painting projects, employee housing (Lakewood) Rowe Apartments, and the Macs' own house. In winter I painted hotel rooms. Some winters I got time off in January, February. The head of maintenance the first few years was a mild-mannered blond guy with a beard, named E.S. (Edmund Shackleford) Lee, who hailed from Ohio. He had been with the Macs since 1972 and worked his way up. He was handy, but self-taught with maintenance. His favorite bands were Talking Heads and A-Ha, and they could always be heard playing in his basement shop. I got along great with E.S.

Mary K. and Marti took office jobs at the hotel. Becki was married and raising a family in the Petoskey area. She returned near the end of my Iroquois time to be in charge of the elaborate flower gardens. Aaron was working in New York City. Sadly, he passed away in the mid -1980s.

When I worked for Jim Francis, we had gotten all the major summer holidays off. So in spring of '84, I took Memorial Day off. When I got back the next day, Mrs. Mac asked me where I had been. "It was Memorial Day, Mrs. Mac. I was off." She replied, "This is the Iroquois, Tommy. We work on Memorial Day."

During my time at Iroquois Hotel, Sam McIntire owned a small yacht, a 26-foot Sea Ray, I believe, which he kept moored to the east side of the hotel's sea wall. But due to unfavorable wave action there, he built a small wooden dock behind the Iroquois Bike Rental, located in the former old State Liquor Store building. The Macs had purchased this when the Liquor Store moved to a new building on Market Street. In the 1960s and 1970s, the Iroquois rented bikes out of the basement at the hotel. When Sam built his new dock, Shepler's were concerned it would interfere with their landing at their first ramp, but it didn't turn out to be the case. Some years the bike shop had trouble keeping bike mechanics, and there were times I was pressed into service to do the work. Sam even thought about sending me to a school for mechanics run by Schwinn.

During my time at the hotel, the Macs built a new house on the beach 1984-85, across from their current house. It was quite unique, and designed by architect Richard Bos. All the rooms were large, but it had only three bedrooms. The outside was natural cedar shake shingles. The trim was cypress, as was the front fence. The den on the east side was all natural varnished wood. Marshall Brothers were the general contractors, and I did some of the painting.

An anecdote I'll always remember from my time at Iroquois is as follows. In early December of 1986, I was enjoying a peaceful day off at home, when the phone rang. It was Mr. Mac. He asked if I could come to the hotel and help him with something. So, I rode my bike down. He had me follow him to the beer storage room. He said, "See all that domestic beer? Put it on a hand truck and take it over to my house." I loaded up several cases of Budweiser, Stroh's and Miller Lite, and stashed them in the storeroom at their new house on the beach. We then returned to the hotel. "Now, see all that imported beer? Put it in a garden cart and take it to your house." So, I filled their cart with several cases of Molson, Heineken, and other imports, and took it home. That's the kind of guy Sam McIntire was.

One night in early March 1988, I got a call around 9:00pm. It was Sam. He asked, "Did I catch you working?" I replied no, and we talked for a while. He sounded a bit off. I found out a couple weeks later, he was suffering from a brain tumor. Mrs. Mac and the girls made sure he got the best treatment possible, mostly in Ann Arbor. In spite of the doctor's best efforts, Sam passed away around the end of November. The service was held at Ste Anne's, Dec. 2nd I believe, and I had never seen so many people there. Star Line ran *Radisson*, bringing all the mainland people who wanted to pay their respects. They docked at The Straits Dock for convenience.

On Christmas, a sprinkler pipe broke at the hotel, on the central section east side of the third floor. It created a big repair project, on what was planned to be a quiet winter. Marshall Brothers sent a crew, Belonga sent Loren Horn as plumber, and I put together a five-man paint crew. Everything was finished in time for opening day 1989 and I continued working at Iroquois through the 1993 season.

Mississippi River Trips

As mentioned in my chapter on ferries, the Star Line ferry company bought their new vessels in Louisiana, and then brought them to Michigan via various waterways. My good friend from the Island, Tom Pfeiffelmann, was the founder of Star Line, and in spring of 1985 he invited me to join the delivery crew for their current boat being built, *Nicolet*. It was the second Star boat to carry this name, but it would not be given a Roman numeral "II" after the name. Other friends of mine, Tim Horn and Tom Benjamin, had previously gone on the delivery trips for *Marquette* and *LaSalle*. Crews were usually ten to twelve members, and five of us left St. Ignace in a station wagon with our gear, early on the morning on June 1st. These were Tom Pfeiffelmann, Captain Mark Brown, Tim Horn, Tom's uncle Lloyd Atwood, and I.

It took us 24 hours to drive to New Orleans, where we met with others at a popular bar called Bronco Billy's. Lloyd decided to keep a written log of the trip. We spent a few hours sightseeing around New Orleans, on Bourbon Street and Canal Street. Being June, and in the south, the weather was hot for most of our trip, temps being in the low 90s. We drove west 100 miles to Morgan City, La. and from there a few miles to Patterson, the location of the (old) Gulf Craft shipyard, in a bayou. Since my trips there, they have built a new yard. The next day we arrived at the shipyard to get *Nicolet* ready for departure. She was still on land, soon to be launched down a short sandy slope, and dockyard workers were still welding steps to the upper deck.

We made a short shopping trip for supplies to a new supermarket chain outlet called Walmart. I was astounded that I got a full shopping cart of stuff for $80. The night before we set sail, ten of us went to a nice seafood restaurant for dinner. The Gulf Craft owners picked up the tab. *Nicolet* carried two new luggage carts, hoisted to the upper deck by a small crane, and some spare propellers on the stern deck.

We left Patterson at dawn, June 4th, via Bayou Teche to the Atchafalaya River. We had built several "bumpers" out of treated 4 by 4s, four feet long, with screw eyes and ropes. The crew would lower them from the top deck when the hull needed protection during docking, or transiting the dozen locks we encountered along the voyage north. Soon, we entered the Mississippi River. Crew shifts were four hours long, from dawn until dusk. The last shift was sometimes a bit longer, in order to make it to the next fuel stop Tom had mapped out. *Nicolet* carried twin 500-gallon fuel tanks. She was powered by two Detroit Diesel 12-cylinder engines with a total of over 1000 horsepower.

Tim Horn recorded the *Nicolet*'s trip on videotape, and I took lots of photos. On my shifts, I was always "chart man," manning a chart, holding a pencil pointing to our current position along the river, for the pilot to see. The total distance of the trip was around 1800 miles, from Patterson, Louisiana to St. Ignace, Michigan, via the winding Mississippi and Illinois Rivers and Lake Michigan. Along the river we frequently encountered "tows" in both directions. There were large river tugs pushing anywhere from one to thirty-six cargo barges each.

Our first night's stop was Natchez, Mississippi, having covered 300 miles. We fueled up, had dinner ashore, and slept aboard *Nicolet*. There was a bar halfway up the bluff at Natchez, called "The Under The Hill Bar." On top of the bluff, a highway led to Natchez proper.

Tim Horn and I took a walk, looking for a market to pick up a few things. We were in our red Star Line jackets, even though it was warm. Two attractive girls our age stopped in their car and offered to pick us up and take us where we were going, but Tim, being engaged at the time, declined the offer.

Our next day's stretch took us past Vicksburg, Mississippi to a planned fuel stop down a long channel to Greenville, Mississippi. However, there was a problem with the pumps, and we continued to an un-planned stop at rural Helena Arkansas, where there was a small fuel facility run by a young backwoods family.

Everyone wore cutoff blue jean shorts, and several little kids played in the swampy beach area, where an old rusted-out tug and another equally rusted boat had come ashore during high water to spend the rest of their days. We didn't spend too much time there, getting what fuel we could before heading to the next planned overnight stop of Memphis, Tennessee. It was a twenty-hour day, and we didn't pull into Memphis until almost 3:00am. Gulf Craft always sent a pilot along on the Star Line delivery trips. This time it was Brian Tibbs, youngest son of Gulf Craft co-owner Scott Tibbs Sr. However, Brian had become ill by Memphis, and decided to take a flight back home. That left Mark

Brown as the only licensed captain, but of course Tom and the rest of us gave him breaks on the wheel. The push tug rigs didn't particularly like small, fast boats on the river, so we got in the habit of radioing them by name, to see if they wished us to check our speed when passing, in either direction. Binoculars were vital aids in scanning the river. There was lots of debris and flotsam on the Mississippi this time of year, and we always had to watch for it. The current was also against us. On the second day of the trip, we hit a small, submerged obstruction that gave us a slight vibration in the port prop, but it wasn't serious. We took on more fuel in the morning, and out third day took us to an overnight stop at Cape Girardeau, Missouri.

Our morning departure times were not always early, since the fuel dock attendants often showed up when they were ready. However, we made over 280 miles a day. Daytime temps finally dipped below the 90s by Cape Girardeau. We left in a morning fog, but day four would clear for us to view St. Louis and the Gateway Arch by 5:00 in the afternoon. There was a floating McDonald's restaurant there on a converted riverboat, but the swift current didn't allow us to dock in one try, much to the dismay of some of the crew. North of St, Louis we passed Alton, Illinois, then on the confluence of the Mississippi and Illinois Rivers near Grafton and Cairo. We took a slight right turn onto the Illinois River. Next came Florence, Ill., and Beardstown. There are seven more locks along the Illinois, which raised us 160 ft to Lake Michigan levels. We transited eleven locks in total since leaving Patterson, Louisiana. Locks were time consuming, especially if we arrived behind a large tow of barges. Those days, we were lucky to make 200 miles. My favorite fuel stop was at the Illinois Valley Yacht Club, "The Ivy Club" near Peoria. It was a hot, sunny afternoon, and many yachts were tied up, and young people were enjoying the pool facilities. There was also a bar and outdoor lounge. I bought an Ivy Club shirt. A large Caterpillar machinery factory dominated the riverside near Peoria.

Our next overnight stop was Ottawa, Illinois, where we visited the Captain's Cove bar. The next day we passed Joliet, and entered the Calumet River where we tied up at Dalton, Ill., southeast of Chicago, and a few miles from Lake Michigan. Along the Calumet we passed the moored freighter *Sparrows Point* and several saltwater ships. We were also lucky to spot the freighter *Cliffs Victory*, tied up and retired from service. In a few months she was due to head overseas in a scrap tow.

We got underway at noon on June 10th from Dalton after fueling up. The three largest skyscrapers on the Chicago skyline were visible for almost 18 miles up Lake Michigan. We drove straight through day and night, making no stops. I was on the wheel for a few hours. We arrived at St. Ignace at dawn, around 5:00am on June. 11. *Nicolet* took Island crew members over to Mackinac around 8:00am, where my mom and girlfriend met us at the Straits Dock.

In spring of 1988, Star Line was ready to buy another boat from Gulf Craft, and Tom Pfeiffelmann put in an order for their largest yet, the 80 ft, 350 passenger *Radisson*. I was again invited to crew for the trip, which spanned May 9-17, and a few of us flew down via Houston, Texas, to New Orleans. There were a few different crew members this time.

Returning crew members were Capt. Mark Brown, Tim Horn, Tom P, Lloyd Atwood, Jay Hoult, and me. Tom's brother, Edward Lee, joined us as a captain, and Gulf Craft's representative was co-owner Billy Peccararo, who had a license. There were about six other crew members. *Radisson* was impressive, with three decks, four engines, and over twice the passenger capacity of the older Star boats. She was the first to have a higher bow.

The night before we left the Gulf Craft shipyard, owner Billy Peccararo threw us a dinner party at his large house near Morgan City. *Radisson*'s trip up the Mississippi and Illinois Rivers was pretty much like the *Nicolet*'s. I again took lots of photos, and was again chart man for my shift. This year we arrived in Memphis coinciding with the "Memphis in May" Barbecue Festival. Hundreds of private BBQs were set up along the riverfront. Ribs, burgers and brats were cooking, with a nearby concert that had thousands of attendees. As I didn't recognize the artist, I asked a couple of ladies who it was. They looked at me as if I were from another planet, and replied, "That's Waylon!" (country star Jennings).

On a particularly narrow stretch of the Illinois, Billy was at the wheel, and *Radisson*, as she always does, was pulling a pretty good-sized wake. We were not checked down, and our wake threw a few small skiffs right up on the beach. When we arrived in St. Ignace, Gene Elmer met us at the dock and the first thing he said to Tom was, "We've got lawsuits."

Every trip, just south of Chicago, we encountered a very old, discontinued, low railroad bridge. This made it necessary for us to remove everything from the pilot house roof to get under it. We also had to flood the bow compartment. As *Radisson* was the highest boat, we also had to man lines and pull her under the bridge by hand, as engaging the engines would have raised the bow slightly. The bridge has since been torn down.

Our last day on the river, in the Chicago area, we found that we didn't have a chart this time for northern Lake Michigan and the Straits of Mackinac. Ed Pfeiffelman told Tom matter-of-factly that he wasn't going around Waugochance Point (west of Mackinaw City) without charts. "Alright," Tom replied. So several of us went to a bar near where we were docked. Capt. Mark Brown and I borrowed a car from two generous women there, with the intention of driving a few blocks to a marina to buy charts. When we got there, the marina did not have the charts, but they told us the next marina, at such-and-such a location would have them. So we drove further around the Chicago area, finding our way to the next marina. Same negative results! We hunted down two more marinas, still not finding the chart we needed. Not one Chicago marina had a Straits of Mackinac chart. Mark and I were lucky we didn't get lost.

We also had to stop and wait at a drawbridge, while the freighter *V. W. Scully* went through. All told, we were gone over two hours and returned with no charts. We apologized to the ladies for having their car so long. Tom was furious at the delay, and chewed us both out, mainly Mark as he was a captain. We left the Chicago area from Dalton, Illinois again, and headed up Lake Michigan.

On the lake, we discovered the radar was not working properly. It was mounted directly on the roof, and the third deck weather curtain was obstructing its sternward scan. It would be necessary to mount the radar on a pole when fitting out in St. Ignace. We pulled into Frankfort, Michigan, where a marina had the charts we needed. While there I spotted the car ferry *City of Milwaukee*, tied up a couple miles away at Elberta.

We arrived in the Straits in daylight and stopped under the Bridge for a brief crew photo. We then headed for St. Ignace at a full speed of 34 mph, provided by the two props and two waterjet engines. Some Star Line crew and some St. Ignace folks came down to the dock to greet us. It was May 17, 1988, and the first jet-drive ferryboat had arrived. *Radisson* would operate on Star Line's new Mackinaw City route.

Star Line Cadillac in St Louis (May 1990)

By spring of 1990, Star Line needed another vessel. Over the winter they had ordered *Cadillac* from Gulf Craft. Tom Pfeiffelmann stopped by the Iroquois Hotel where I was working to let me know of the upcoming trip. The McIntire ladies weren't too happy that I would be leaving again during the peak time for "getting ready to open" and they made Tom wait in the lobby for several minutes while Mary Kay went to get me. The trip would be April 30 to May 7. Tom himself would not be able to ride up with the *Cadillac* crew on the delivery this time, and he put stockholder and veteran crew member, Ken Karsten, in charge. *Cadillac* was the first Star Line vessel equipped at the shipyard with a v6 center engine, powering a small waterjet angled up at 45 degrees: the first "rooster-tail" in their fleet. She was 65 by 22 feet as *Marquette* and *LaSalle* had been, but she had the high bow of *Radisson*. She would also sported the largest pilot house in their fleet, 9 by 12 feet, but with a ceiling height of only 6'3." Rumor was that the pilot house was something Gulf Craft had "lying around" the yard, possibly planned for a different vessel. *Cadillac* was the first Star Line boat to have cabin heat.

Captains for this trip wee Mark Brown and Ed Pfeiffelman again, and the shipyard captain was Scott Tibbs, eldest son of one of the Gulf Craft owners. As usual, I was to the charts on my shift, in the spacious wheelhouse. The trip proved to be pretty much incident free, but we did get turned around in heavy fog once on the Mississippi, coming out of a channel after fueling. But we quickly corrected course when we noticed the current was going in the opposite direction to what we'd expected.

Cadillac was the only Star Line vessel equipped with a deluxe horn, a melodic, triple trumpet brass Kahlenberg horn, costing $5k in 1990 prices. Gulf Craft co-owner ,Billy Peccararo, had told Tom during his brief time inspecting *Cadillac*, "Don't expect to get one of those every time."

It was necessary to keep the center, "jet" engine running at all times, as it powered the generator. It was just a case of knowing when to crank down the big roostertail. Once while waiting for a lock, the main 12v71 engines were in neutral and the jet was engaged. We were creeping forward at about one mph. South of Chicago, we went under a highway bridge with the jet cranked up. It sprayed over the bridge, but fortunately there was no traffic.

We arrived in St. Ignace during daylight, and a lot of people on shore stopped and gawked at their first site of a Star Line roostertail. *Huron* was pulling out of the Arnold dock as we passed by, and we exchanged salutes. It was May 7.

I had one more opportunity to go on a Star Line delivery trip, in 1993 for the *Joliet*. However, I chose to decline, not wanting to press my luck again with the McIntires and the Iroquois. By my estimate, I had logged around 5400 miles of river and lake travel on my three trips from Morgan City to St. Ignace. They were great experiences, always to be remembered.

A Lifelong Love of Music

My first taste of music came when I was around 9 years old. I had heard Johnny Horton (Sink the Bismarck, Battle of New Orleans), and David Seville (Witch Doctor). on the radio. I persuaded my mom to buy me the 45. Other than that, I had heard some of mom's big band music (Glenn Miller) on our record player. Mom played piano and organ, and had played clarinet in high school in Alpena. She was also an excellent singer, and sang in the Ste Anne's choir with Mary Kate McGreevy, Gloria Davenport, and Nova Therrien. My dad was not musically inclined. Neither was stepdad Ray, but he did dabble on the organ.

The first album I bought, at a St. Ignace grocery store when I was 13, was Herb Alpert and the Tijuana Brass: Whipped Cream and Other Delights. I liked the sound of horns, and had heard Uncle Dennis playing a bit of Herb and the Brass. Then of course there was the album cover...a 20-something girl covered entirely in whipped cream (it was actually shaving cream). So many males bought it just for the cover, but I played the heck out of it, on my own new record player. Also in 1966, my friend Tim Horn turned me onto a new album of his by the instrumental band, The Ventures. The title was "Batman Theme." It contained 12 spy-oriented tracks. The *Batman* TV show, starring Adam West, was very popular that year among young people. I didn't really listen to radio except for WIDG-AM in St. Ignace, sporadically. In 1967 WJML-FM in Petoskey started featuring 90 minutes of rock every day after school, from 3:30 to 5:00. I liked tuning that in and hearing all the current new rock songs.

The nice thing about being a teen in the 1960s was to be in on the ground floor of the rock revolution in music. I was hearing all The Beatles songs when they were brand new hits, and many others. I had seen The Beatles once on an Ed Sullivan TV show appearance, doing I Want To Hold Your Hand. They were just OK to me. I missed their other Sullivan show appearance, because I was watching "The Scarecrow of Romney Marsh" on Walt Disney. In those days, most people only had black-and-white TVs with the roundish screen. The first color TV I saw (still roundish) was in seventh grade, when

Superintendent Byron Antcliff invited a bunch on the boys to his house to watch a U of M football game. He and his wife lived in the log cabin (now on Mahoney), which back then was on the lot where Wayne Zwolinski's house would be built, near the school. It's now Callewaert's. In 1968, I was introduced to more rock albums, such as The Rolling Stones, by my friend Jim Dankowski.

Jimmy also turned me onto a unique album called "The Zodiac-Cosmic Sounds." It had 12 tracks, one for each sign of the zodiac. A narrator recited prose over each instrumental track of mostly psychedelic music. Studio musicians did the playing. I still have my copy of the LP.

In 1966 my mom and Ray bought me my first guitar, a Kay sunburst acoustic. It was nice to have, but hard on the fingers for a beginner. Nevertheless, I learned some songs, but decided I needed an electric ASAP. On a shopping trip to Cheboygan in 1967, I bought a red sunburst Kingston electric for $60. It was off-brand, but a decent entry level electric. Also in 1967 I saw the excellent St. Ignace band Gross National Product perform at the Community Hall. They did great versions of all the current rock hits and album cuts. I decided then and there I wanted to be in a rock band. Their singer was local AM radio DJ, Joe Arthur. Band members were Jim Fitzpatrick, Steve LaJoice, Billy Becker and Louis Spencley. In May of 1968 they played for our MIHS prom, which I was excited about. I also saw them play at a Battle of the Bands in St. Ignace. While attending LaSalle HS for a couple months in the fall of 1968, I met Louis' kid brother, Rodney, who was a drummer as Louis was. In May of 1969, Rodney's band was hired to play the Island prom, and they needed a fourth member, so they asked me to play the gig. Other members of Rod's band were guitarists Henry Smith and Randy McLean. After that I began work on my own band, as covered in another chapter.

By 1967, in addition to listening to The Ventures, I was into Paul Revere and the Raiders, The Standells, The Outsiders, and Billy Strange. I continued to build an album collection. Every time my family went to Cheboygan or Petoskey, I bought something. I was also in a mail order record club (who wasn't?). By 1968, I was getting into Michigan rock bands, The Amboy Dukes, Frost and SRC. When I went to college at NCMC in Petoskey, I haunted the record store every week, and bought something whenever money allowed. My stereo systems were getting better along the way. The year after graduating

NCMC, I went to Western Michigan in Kalamazoo. There were several large record stores there. One was even right near campus. In addition to buying albums by my favorite acts such as Alice Cooper and Deep Purple, I also sought out new music and rare progressive rock stuff. I recall buying the first Queen and the first Kiss albums when they hit the shelves. At WMU, my friend Rich Schroeder introduced me to a lot of new music.

Another group I had gotten in on the ground floor with was Uriah Heep, their first and second albums in 1971, while living at Silver Birches. I bought Journey's first album when they were a jazz-rock spinoff of Santana, and at WMU I discovered Styx's first album, before they ever had a hit.

My progressive rock favorites will always be Yes, King Crimson, and Starcastle.

In addition to albums, I also had a few homemade reel to reel tapes and cassettes.

In the 1980s, I continued to grow my album collection, picking up on the New Wave movement. The Cars, Joe Jackson, Letters to Cleo, and Blondie were particular favorites. My albums were now joined on the shelf by compact discs. When the alternative rock trend arrived in the 1990s, I embraced it, too. I had a female summer employee friend who introduced me to a lot of it. I also got into the indie music of the late 1990s and 2000s. In recent years, my favorite bands are Warpaint, St. Vincent, Pacifica, and Khruangbin. I was never much of a fan of country or classical music. My collection eventually topped out at over 800 albums and 500 CDs. This was augmented when I became caretaker of Michael Hart's and Pat Granning's LP collections.

In the 2000s it's nice to see LP albums making a comeback, and I still purchase them.

Golf, 1967-2001

I come from a long line of golfers on my mother's side of the family, and as with many Island kids, I took the sport up at a young age. My grandfather, Phil Dufina, had been an excellent golfer, once shooting a 29 for nine holes at the Grand Hotel course. He was golf pro there for a few years. His brother, Jim, followed him in that job. All of the Dufina brothers of that generation were golfers. My mom's brother, Dennis, was also an avid golfer, and was golf pro first at Wawashkamo and then at The Grand. One year, his assistant pro at The Grand course was future long-time dockporter Glenn Woulffe.

I worked for Dennis in 1967 at Wawashkamo, and later briefly in 1979 at The Grand Hotel course. I preferred golfing at Wawashkamo, but I would also play the Grand. I was never very good, basically being a "bogie" golfer, which is one over par per hole. My best rounds ever were 39 at the Grand course and 40 at Wawashkamo. My forte was putting. When it became time to buy a new set of clubs for my adult golf life, my Uncle Dennis picked me out a nice set of mid-level Wilson Johnny Miller's. I never liked using the driver (1-Wood), even when my dad later got me a custom driver. Instead, I drove with the 3-Wood, which, combined with my three-quarters swing, meant I never had long drives. I also liked using a 5 and 7-Wood. My putter was a PGA mallet style.

My usual foursome was with Armin and Matt Porter and Michael Hart, but I also played with many others over the years, including Clark Bloswick, Herbie St. Onge, and Louie Bunker. Clark recalls how following a bad drive off the tee, Armin often threw a club, sending it spinning through the air like a helicopter blade. No one cared, though, since he was a Member at Wawashkamo. I joined the golf leagues when they formed at Wawashkamo and was with them for many years. When the league moved to The Grand one year, I played there also. I also played in a few off-Island tourneys at Alpena and Petoskey. I also enjoyed playing in Florida during the winters of 1969-70.

There was a nice 18 Hole Par-3 course near our house called Glen Oaks. I played it a lot, often with my cousins, Bob and David Dufina. I joined the golf team at Clearwater Central Catholic High School. As a prank, my cousins told the coach was a very good golfer, so he made me "1st golfer" on the team. There were four team members and two alternates. When I kept performing badly against other schools' lead golfers at matches, Coach Jim Smith was not amused, but he assumed my game would come around, which of course it never did. Smith was also football coach and a teacher. At the sports awards banquet at the end of the year he called two of us aside. "Chambers, Kleinfeld, you guys don't get your golf letter, since you never turned in your birth certificates." I think this gave him some smug satisfaction. I was disappointed. Playing on the beautiful Florida golf courses had been quite the experience, though. One of my favorites had been up the coast at Tarpon Springs.

One time on the Island in the 1980s, on my way to League at Wawashkamo, I was cycling down Garrison Road at speed with my golf bag across my basket. A group of saddle horses came off a side trail. They were from a stable that only operated a few years. One horse shied and kicked me, sending me into the woods. Another golfer came along and saw the mishap, continued on to the course and called an ambulance. Matt Porter and Eddie Horn came back from the course to check on me. I missed the day's match of course, and received a few cuts, a slight concussion and a black eye.

Later, a new golf course opened on the Island, "The Woods," owned by Grand Hotel. It was across State Road from Wawashkamo, north of Stonecliffe. I tried it a few times, but the fairways were narrow, and not kind to a golfer who often sliced his drives, like me.

I had always taken a six-pack of Budweiser with me when I golfed. Not to get drunk, but just to relax my game and "loosen up." After I quit drinking, I played once at Wawashkamo in summer of 2001, but didn't score well. I decided to give the game a rest. The following summer, my set of clubs which were stored at the course, came up missing. That pretty much sealed it, then. Mark Bunker even offered to give me a set of clubs. I thanked him for his generous offer, but my golfing days had come to an end, at age 48.

Playing Games

When I was a child, I played the usual board games, as many youngsters do: Parcheesi, Sorry, and Monopoly. I quickly graduated to chess, which I learned at 7 years of age. My Uncle Dennis taught me the winter we were living out west. He made me a homemade chess set with cardboard pieces on top of bases made of Popsicle sticks cut to length. I was soon able to buy two nice hand-carved wooden chess sets, in Nogales and Juarez Mexico, at bargain prices. I still have them to this day. I got a nice Renaissance set when I was in college. Chess became my game of choice for the next 15 years, playing a hundred or more games a year. My most frequent opponent was Tim Horn, a close friend from the Island. I once lost a game in HS to exchange student Brian Lu, which Tim ribbed me about, but Tim lost his share of games, too. Dan Johnson became another frequent chess opponent. I would play anybody, it didn't matter to me. I also had chess matches with some of the Catholic priests, who rotated through Ste. Anne's. Fr. William Schick gave me a chess book.

Sometime by early in my senior year in Florida, I had a winning streak of over sixty games. But being a chess shark is like being a gunfighter in the Old West... there is always somebody better. I recall losing a game at CCC to teacher and baseball coach, John Ash, an older guy with a Deep South accent. I continued to play in college, at NCMC and WMU. Scott Sweeney was the best player at NCMC. At Western, my opponents were good friends, Tom Stieber, Steve Coombs, and Rick Thompson. In the late 70s, I played a few games against my landlord's son, David. He was very good, would play blindfolded, and always won. I continued to play chess through the 1980s, though not as often. My last game was in 2000, against a cook on the Island named Big Ed.

The 1980s were a good decade for playing poker on the Island. Nothing extravagant, just quarter to $1 ante. There was a group of half a dozen of us who rotated games between one another's house—Rick Wessel, Clay Fuller, George Bodwin Jr., Mark Bunker and me, to name a few. I remember several games at Johnny Fisher's house, at one

Playing a board game (1985)

of which businessmen Dennis Brodeur and Ron Dufina showed up. The biggest loss or win was usually around $100.

Darts, ping pong and croquet each went through a period of popularity with my circle of Island friends. I learned cribbage at an early age, having been taught by my grandma, Beth Roussin. I enjoyed it for many years, playing quite a bit in junior college, mostly with Rod and Ned Taylor. In the 1980s and 90s my steady opponent on the Island was Clark Bloswick.

In 1974, I was introduced to what would become by favorite type of game system for over twenty-five years...board "war-games." My friend, Michael Hart, showed me the first one, for the battle of Gettysburg. Avalon Hill and Simulation Publications Inc. (SPI) were the primary companies produced games for pretty much any battle, war or conflict you can imagine. Dice were used to resolve battle odds.

This also led to me meeting a fellow gamer, Armin Porter, eldest son of a West Bluff family. We developed a close, lifelong friendship. I can't begin to count the number of hours I spent between 1974 and 2000 playing war board games. The three of us purchased games, and my collection eventually numbered over 100 different simulations. We were the main three players, but others in our circle of friends occasionally joined in, including Matt Porter, Steve Goldsmith, Nick Haan, Steve Schneider, and Steve Beaver. There were even all-night games sessions, between four and five players, mostly when I lived at Haan's. I remember Stephen Doud stopping by to observe one such game when he lived at Reid House, and was going through a divorce. From the late 1990s to 2003, Armin, Mike and I played the desktop computer versions of war games. There were even solo computer games available, which I played until around 2008.

A board game similar to war games, which saw a lot of play from the late 1980s until 2005, was Rail Baron. It was best with four or five players. It used dice and money. Players traveled the country amassing a network of connecting rail lines, and building a fortune. Another good board game Michael Hart found at Alford's was Great Lakes Cargo. It's very hard to find nowadays. We also resurrected the childhood game, Scrabble, and had many spirited four-way games either at my place or Lakewood cottage. Anne Bronfman was the owner then, and Mike was caretaker.

A strategy game we played in 1977-79 while I was living at Haan's was APBA NFL Football. It had packets of forty player cards for each

NFL team, several large charts, and dice. You had to coach against your opponent's team in a simulated game. A bystander who knew the game would often volunteer to read the charts. This was in the pre-computer game age. Players were Mike, Armin and me, along with Phil Porter and John Croghan.

A Return To Music, 2008-2022

I had not actively been in a band since Morningstar broke up. I still had my Mosrite electric, and I bought my first electric bass, also a solid body Mosrite, in June 2003. I got it from Lou Lacey of the Grand Hotel orchestra. I was aware for years that he had a Mosrite, and when he decided to part with it, I jumped at the chance. That began my years of collecting bass guitars. I had figured out years ago I wasn't a very good guitarist, and if I was ever going to do anything musically again, I'd better learn bass. Over the next fifteen years, I had collected twenty-eight bass and thirteen electric guitars.

In late Nov. of 2008, my buddy, Jason St. Onge, contacted me. He was putting together an ad hoc band to play Christmas Bazaar weekend at Patrick Sinclair's Irish Pub.

Normally, he would have had his regular band, Tricky Dicky & the Spoonmen do the gig, but singer Paul BeDour was downstate for his girlfriend's birthday. I had sat in with Tricky Dicky a couple times in 2006-07. He asked Myk Rise to play guitar and sing, and me to play bass. Jason would be the drummer. Sinclair's was the only game in town this Christmas Bazaar, as the Mustang was closed for remodeling. Sinclair's was located in the former Ty's Restaurant location, on Main Street. We had decided on about thirty classic rock songs, which we all knew. Jason dubbed the band "Armin's Army" in honor of our mutual friend, Armin Porter. The place filled up fast, as there had not been a bar beside the Mustang open since Halloween. Sinclair's was packed for our gig, probably 200 people. Everyone wanted to dance, and Jimmy Fisher recorded it on his phone. The gig was a great success, and I wondered to myself in amazement if this was what music performance was going to be like now.

Armin's Army decided to stay together for limited performances, our next one being on the outdoor patio of the Chippewa Hotel's Pink Pony bar. Our namesake, Armin Porter, would even come to the shows, listen, and have a Stroh's beer. Myk Rise had a regular band also, as well as being the featured solo musician at Pink Pony every summer.

Still, Armin's Army managed to fit in a handful of gigs over the next few years. Among my favorites were at the Island's now-renovated Community Hall, for three Januarys, around 2010. Jason and I came up with the idea of having a yearly Island "Heritage Day" two weeks after New Year's. We pitched the idea to Mayor Margaret Doud, and she liked it. The late afternoon/evening event consisted of Islanders bringing their old photo albums, a slideshow by Robin Dorman, written trivia quiz with questions I chose, refreshments, music, and dancing. About 100 locals came.

For the 2011 season, Myk Rise decided to bring in a different drummer for our gigs. Donny Sorenson of St. Ignace joined us for shows at the Pink Pony Patio and Huron Street Pub. In 2012, Cheboygan singer/songwriter Charlie Reager started drumming for us. He had prior band experience. Also in 2012, I came up with a new name for the group, "Chimney Rock band." Chimney Rock is the old school name for Mackinac landmark Sunset Rock, on a bluff on the Island's west side. When the Sip N' Sail cruises started up, we signed on for gigs, and became one of their early acts. We usually played three cruises a summer for about five years. While Charlie was drumming for us, we picked up two off-Island gigs. We were invited to play the Cheboygan Music Festival in July of 2015, with three other acts. We also played an ice festival on Cheboygan's Mullett Lake. Another musician, Nate King, was coordinator and would help us set up. We also continued to play Pink Pony gigs, either on the patio or for opening night in early May. During one patio gig, my old drummer, Matt Finstrom, was visiting town, and sat in with Myk and me for a few songs.

Occasionally, we did winter and mid-winter gigs at The Mustang. Tony Brodeur was running it. The Christmas Bazaar was always a popular weekend for bars to hire bands, and after that successful night back in 2008, I tried playing them every year. When The Mustang reopened after remodeling, I booked Bazaar weekend there for Chimney Rock band.

Charlie drummed it with us one year. In around 2016, Myk asked his old drummer from the Myk Rise Band, Bill Newland, to start gigging with us. Charlie's wife had just had a new baby and he didn't want to stay overnight on the Island. I enjoyed working with Bill for the next few years. He was also our drummer when we started performing as an opening act for "Music In The Park" for three years,

Chimney Rock band (08/20/2017)

and for one-off gigs at Mission Point, The Gate House, and a 60s party for Wawashkamo Golf Course held at the Yacht Club. Phil Porter was emcee

I also did three gigs with singer-songwriter Mary McGuire over the years. One was a New Year's gig at Patrick Sinclair's Irish Pub with Gary Rasmussen and Nino Dmytryszyn joining Mary. Sinclair's even gave us dinner. She and I also had one Christmas Bazaar gig at the Mustang, with Jason St.Onge and Paul Bedour, although Jason got called away early to respond to a plane crash with the MIFD. In addition, Mary and I were in one of the bands playing a summer gig at The Mustang to honor Dennis Brodeur.

Other one-time gigs I did were with Jason, Paul Bedour, and Brian Thomas at Horns Bar one spring. I also took part in the Thursday night summer jams at Patrick Sinclair's Irish Pub one year. The band of Myk Rise, Chris Tamlyn, Nino Dmytryszyn and I was well received. Another drummer I worked with at Sinclair's was Justin Bartels, bar manager from the Iroquois. Sometimes Paul Bedour also sang with us there.

I was happy when Annie Porter invited me to play at her wedding reception at Grand Hotel, joining Myk Rise and musicians Bill

Newland, Jason Haske, and Nate King. This meant wearing my best suit and tie, of course.

Some singers who sat in with our Chimney Rock band included Adrienne Rilenge and Carolyn May.

We didn't have and performances in 2020, the year of the Covid-19 virus. My last gig with Chimney Rock band, at the time of this writing anyway, was at The Mustang, for Christmas Bazaar weekend in December 2022. The lineup was Myk Rise, Bill Newland and I.

Is Mackinac Haunted?

I don't like to be a party to tabloid stuff or sensationalism, but no coverage of Mackinac Island would be complete without exploring the topic of haunting and haunted houses. A side-effect of claiming something is "haunted" is that it can give that place a bad reputation. I'm not a fan of the haunted Mackinac pages you have no doubt seen already on social media. Most of their claims are bogus. One such claim is "The Drowning Pool" on the beach near Mission Point. It was just beach shoreline until MRA/Mackinac College excavated and dredged it for a possible marina site. The marina project never came to be. Nowadays, it's a simple cove.

The biggest example to me of a true haunted house was the old Harvey house, on the bluff between Stonecliffe and Hedgecliffe. It was dark, three-story. and run down. It had a separate outbuilding in the rear, an icehouse type, with a cellar. When I was a teen, my friends and I used to ride out there to check it out, but never at night. There was an old piano in the living room. The house had lots of broken windows, and the whole place had a creepy feel. There was some furniture but no real belongings. Strangely, it had not been vandalized. I never remember it being occupied, and the Harveys must have moved out by the mid-1950s. In 1980, lightning struck the house and it burned to the ground. An empty, grassy lot sits on the location now. The ice house remained for a few years before being torn down.

Another place I considered to be haunted was the big lodge at Silver Birches. When my band lived out there in 1971, in the rear cottage, we used to practice in the big lodge, and do laundry there. We kept the doors locked. We were constantly finding objects moved around, out of place, and not as we had left them. One time a microphone disappeared, but not its cord. Who would steal a microphone without also taking the cord?

Another candidate for a haunted house was the Bailey House. An elderly man died there in the turret room, a victim of a fire. My list goes on to include Small Point, Pine Cottage, and the Dufina cottage in the Annex.

The abandoned 'Haunted' Harvey House. Photo by Loren Horn

Mackinac Island Trivia

When the young kids from the east end were playing outdoors in the 1950s and 60s, their mothers would often tell them, "You have to come home when the streetlights come on."

~ ~ ~

For a few years in the 1980s there was a top-notch Mexican restaurant in town. Frank Nephew had decided to remodel the Pickle Barrel in 1982, and created "Everybody's Little Mexico." The decor was great, and it would become known for authentic Mexican cuisine and drinks. My buddy Armin Porter was manager/bartender. Severs included Jim Balone, who is an author nowadays, Charley Hegarty from the East Bluff, and my girlfriend at the time, Carole. Sadly, the restaurant was destroyed in the LaSalle fire of 1989, and not reopened in the new Lilac Tree building.

~ ~ ~

There were three dray lines in town for people to choose from when I was growing up. They had not been consolidated into one service by Carriage Tours yet. Cowell's had the red drays. Fuller Cowell was the patriarch, and sons George, Johnny, and Reuben helped out. Sonny Therrien, who had the US Mail contract, owned the gray drays. His brother, Mike, drove the second dray when it was busy. Milton Bazinau had the green drays—Milt's Dray Line. I think his brothers, Randy and Wally, sometimes drove for Milt. In around 1970, Milt sold his business to a new fellow in town named McNabb. Grand Hotel had its own dray, mostly for luggage. Steve Kosar drove it for around twenty years. The State Park also had their own dray, as did MRA/Mackinac College.

~ ~ ~

Three positions on the Island changed leadership frequently during my younger years: police chief, school superintendent, and doctor, at least after Dr. Solomon anyway.

~ ~ ~

Although Doud's has long been known as the only market in town, there have been other places to buy groceries over the years. Harry Ryba operated a general store/party store for over a decade, where

Mary's Bistro is now. I recall a guy named Leo Suttles managing it one year. In 1964 - 65, Mary and Ray Summerfield had a small party store in a building behind Bay View, about where Watercolor Cafe's outdoor seating is now. It did a good summer business with people from the east end of town and East Bluff, and also people waiting for a Straits Transit ferry. The east wing of the new yacht dock hadn't yet been completed, otherwise business would have been even more thriving. The Davis Store downtown also carried canned goods and dry goods in the 1940s and 50s. Also, there is the Harrisonville General store.

~ ~ ~

Coal for the Coal Dock was delivered by a Wyandotte Chemicals self-unloading freighter, from about 1909-1970. The vessels got larger over the years. They tied up at the Arnold Dock, and swung the boom over. I also remember seeing one photo of an Oglebay-Norton freighter pulling in.

~ ~ ~

Henry Trail, from Fort Holmes road to Skull Cave, used to be called "Suicide Hill." It was on the real old Carriage Tour route in the 1950s and early 60s. The thinking was, it was "suicide" to take your buggy down this hill, especially without brakes. Some of the older drivers recall taking it, but only if passengers had gotten off at the fort.

~ ~ ~

A Mackinac Island pastime of the 1940s through 1980s was "Blacksnaking." As winter was giving way to spring and a sunny unseasonably warm day came, a few young Islanders would gather on the slope in front of Fort Mackinac's West Blockhouse. They each brought a bottle of cheap wine or a six-pack of beer, and consume it while soaking up the afternoon sunshine. This activity was enjoyed by town kids and Harrisonville kids alike. In recent decades it hasn't been such a tradition, but it still occasionally happens.

Considerably fewer places served alcohol when I was growing up than today. By my count, there were nine then, most of them very popular with the locals. There was Mary's Pantry (Gate House) run by the Bloomfield family. Ed Horn owned the Palm Café (Horn's). It had a neon palm tree sign hanging outside. Hardy's on Astor Street catered mostly to black clientele. It was run by Washington Hardy. The location is now The Mustang. The Pink Pony was still there in the old days, run by Nathan Shayne. The old Village Inn, near the corner of

Main and Hoban, was run by Bob and Arlene Chambers, who in turn sold it to Dennis Brodeur and Jack Chambers. The Grand Hotel and Snack Bar of course served alcohol. There was a small bar in the east end of the Lake View Hotel, owned by Jim and Mary Cable. Iroquois Hotel's Carriage House restaurant and lounge opened in 1960. John and Ethel Ross opened the Lamplighter in the Murray Hotel around 1965. I have also heard stories of a place, maybe 100 years ago, called the "Bucket of Blood." It was a saloon located about at the head of the modern day Shepler's Dock. Currently, I have lost count of all the watering holes, but there are over twenty.

~ ~ ~

The Borden Milk company sent an enclosed milk wagon to the Island for a few years, around 1962-66. It was pulled by a single horse, and the company logo was on the side of the wagon. It loaded its product at the Straits Transit dock, and made its way around town with deliveries. I think their milk came from Cheboygan. I knew the driver in 1965, a young guy named Bob. He kept his Schwinn Varsity 10-speed in the back of the wagon also.

~ ~ ~

The Island's boardwalk, which stretches from the Iroquois Hotel to just east of the Grand Hotel's swimming pool, was first built in 1942, financed by a tourist tax by one of the boat lines. Before it was built, there was just a sloping bank down to the beach and water. Around the turn of the 20th Century, Grand Hotel had a long wooden dock on the beach below their pool. The boardwalk was of all wood construction, both deck planks and the posts and stringers. The deck planks ran straight across the supporting stringers.

The boardwalk lasted about twenty years before it fell into disrepair. W. Stewart Woodfill of Grand Hotel took the initiative to have a new one built. He recruited Dr. Eugene Petersen of the State Park to help facilitate the rebuild, as the City wasn't interested. After experimenting with southern yellow pine, the builders settled on Douglas fir for construction material. The rebuild was financed through a revenue bond program. This time it had a chevron (herringbone) pattern to the walkway. The State Park maintained it, and its next replacement came in 1986. I remember local man, Johnny Ray Gallagher, of Harrisonville and ex-dockporter, Sam Oliver, nailing down the 2x8 planks with 20 penny nails and 2-lb sledges. I was working at the Iroquois at the time. The boardwalk seems to last twenty to twenty-five years before it needs major work again.

~ ~ ~

There once was a deer park on Mackinac Island. Around a dozen were kept in a wire-fenced enclosure of a few acres, between Garrison Road and Custer Road, running as far as the south end of Ste Anne's Cemetery. In the mid-1940s it was discontinued, and the deer turned loose, the State Park citing lack of financing as the reason.

~ ~ ~

In 1977, there was a World War II museum here, located one door east of The Big Store. It was run by John Barrett of suburban Detroit, who loved Mackinac Island, and was a WWII buff. It had glass display cases with mannequins in uniforms, weaponry, flags, medals, and documents. He even had a German jeep (kubelwagen). All the nations participating in WWII were represented. The museum was open the whole season, but as one might expect, it didn't do a great business. I went in multiple times and checked everything out. At the end of the season, John sold his collection to the Confederate Air Force, a museum in Texas.

~ ~ ~

Tennis and shuffleboard courts predate my time, but I felt they were worth including. Tennis has been on Mackinac since just before the turn of the 20th Century. The earliest courts were, the Grand Hotel of course, and also at the Henry Leman (Hamady) cottage on the West Bluff. The Grand's first two courts date to circa 1900, and were originally of grass, and located where the tea garden is now. Around 1920 they were moved to their current location near the road to the swimming pool. Their number was increased to four, and they received a clay surface. Two shuffle-board courts were built near where the Woodfill Memorial is now. The other old tennis court was at the Henry Leman cottage, across the road, in the large lot near the carriage house. It had a grass surface, and was fenced. A large grape vine grew up the side of the fence. It was discontinued in the mid-1990s, and it has remained a big grassy lot to this day. In the late 1950s, Moral Re-Armament built a tennis court on the waterfront, about 100 yard west of Robinson's Folly. It now has an asphalt surface, but may have originally been grass or clay.

In the 1940s, two shuffleboard courts were built near the yacht dock, just east of the Coast Guard Station. In summer of 1974, Annex resident, Gerald Ross, cleared land and built a new asphalt tennis court on the plateau down the slope from his cottage. It's now Hahn's. I helped work on the project, with contractor Ray Summerfield. Lastly, the Mackinac Island State Park cleared land across Garrison Road for a new asphalt tennis court in the fall of 1976, built the following summer, 1977.

In the 21st Century, the Grand's two remaining tennis courts are now a green synthetic modular material, with small holes in the connecting tiles.

~ ~ ~

There was a tradition on Mackinac when I was younger where the Mayor would request that all businesses close, and all work cease, from 12 noon until 3:00pm on the Christian holiday of Good Friday, just before Easter. It was a solemn observance and lasted for many decades. Then, in the early 1990s, a bar owner and a hotel maintenance man decided to buck the system. The bar stayed open and sounds of a power saw could be heard from the hotel. Just like that, the somber tradition was dead. Business places never closed again on Good Friday. Both of the men responsible have passed away several years ago. It would be a nice small-town tradition to return to, but in the 2020s I don't think many establishments and crews would be keen to observe it. They would say they're simply "too busy" getting ready to open for the looming tourist season. And what would the construction crews do? Get back on the midday ferry? It's just another example of change in a small town.

~ ~ ~

Fire protection on Mackinac Island, as with most communities, advanced through the decades. We originally had an old horse-drawn, coal fired steam fire engine, with a prominent smokestack. When I was a kid, it was always a featured attraction in the Lilac Day Parade. It now resides in the Surrey Hills carriage museum for all to still see. The fire department was, and still is, all volunteer, drawn from the local residents. The first motorized fire truck appeared in the 1930s, and new, updated ones were purchased by the City about every ten years. It was housed in a garage on Market Street, behind City Hall. In the 1950s and 60s, the Island actually had three fire truck crews. The State Park had their own, manned by State Park employees, who were often

Islanders. Their truck was kept up the hill, in the State garage. MRA also had their own fire truck and crew, which they felt was necessary for quick response in the event of a fire at one of their structures, but it also responded to fires within the city and environs.

In recent years, the City acquired another fire vehicle stationed up the hill, near Harrisonville. Notable fire chiefs I recall through my life have been Armand "Smi" Horn, Dennis Bradley, and Jason St. Onge. Major fires of the past included Doud's store at the head of the Arnold Dock, which burned down in the winter of 1943. Another was the Sherman cottage on the West Bluff, which burned down in Dec. 1978. Lightning strikes were responsible for two fires that resulted in total loss: the old abandoned Harvey cottage between Stonecliffe and Hedgecliffe, around 1980, and the unused barn at the old Helen Wagner property (now McCabe's) at British Landing in the 1990s. The Wagner house had also burned down, in September 1967. It had been owned by Ray Summerfield. I had been living there, and was at a school dance at the time of the fire.

There was a fire at Chippewa Hotel around 1963, which was put out before it became severe. Deaths due to fire included 90-year-old Nellie Doud in 1964, in the house just west of Woodfill's, and Bob Bailey around 1970 in his house on Church Street. Sheldon Roots, curator at the Stewart House, and his sister, Beth, died in a propane gas explosion in the mid-80s in their house near Mission Point. There was a fire at The Big Store in 1987. By far the largest fire on Mackinac Island was that of the LaSalle Building on Main Street, in Sept. 1989. The second floor had been employee housing, with several shops on the main level. A young couple died of smoke inhalation in the fire, and La Salle was a total loss. The shops were destroyed by water damage. This tragic fire brought about much needed upgrades in all Island employees housing, as well as a surge in the installation of sprinkler and alarm systems in hotels, businesses, and housing.

~ ~ ~

Over the years I have had a like/dislike relationship with the Mackinac Island State Park. A lot of their projects, as I saw them, were of little benefit to tourists, and certainly not the Islanders, but rather thought up by some bureaucrat in Lansing. Certainly they did many good things, such as creating paved North and South Bicycle Trails, banning burning, and obviously the wonderful restoration of Fort Mackinac, since 1957. Among their real headshaking projects I include the lengthy and costly tearing out of the hedge behind Marquette Park, only to replace it with another hedge that was slightly tighter knit.

I was also dismayed when the State chose to cut down all the Christmas Trees in Soldier's Garden. One thing they thankfully discontinued, was oiling the roads, to keep the dust down. They have had a longtime policy of spreading imported gravel on all the dirt roads. Sure, dirt roads get muddy, but the gravel they use is no fun to bicycle on.

Road cover was taken to the extreme in the 2010s, when they tried covering Scott's Cave (Shore) Road with golf-ball sized pulverized asphalt, which had been torn up during M-185 repair. Try riding a bike on that surface! Scott's had always been a quiet, beautiful and pleasant trail to cycle along. As of this writing, the State is currently building a Milliken Nature Center at Arch Rock. I think William Milliken was one of our best State Governors, but this structure is scant tribute. The architecture is totally out of place for Mackinac Island. It looks more like a rest stop along I-75. Who signed off on this design anyway? It should have been Colonial in style, maybe resembling the Biddle House.

~ ~ ~

A habit on the Island from bygone days was having private burn barrels to burn trash. Folks used a 55-gallon drum and cut one end off it. I also took advantage of them on the two rental properties I lived in, from 1982-95. I only burned paper products, but some people took it to the extreme and burned garbage, copper and things that didn't really burn. Of course, this created unneeded smoke and a smell in some neighborhoods. There was even an "old burner" on a large lot and old dump west of where Iroquois housing is now.

In the 1980s the State banned burning in the State Park, and they tried to limit open burning at the City landfill. Some residents soon came out against burning also. Burning decreased as compost bags became more popular. Finally, The City banned burn barrels in the late 1990s. A few die-hards kept using them until around 2005.

~ ~ ~

Mackinac's favorite burger place, Mighty Mac Hamburgers, was started in 1969 by Frank Nephew. Prior to that, Ling Horn and Ruth Clark managed Tom Shamy Linens. For many of the early years it had "flame broiled" burgers, cooked on a conveyor belt. This was replaced by a conventional grill in the early 1990s. Managers who ran Mighty

Mac the longest were Doug Nephew in the 1970s and 80s, and Armin Porter, from 1990 until his passing in 2013.

~ ~ ~

Since the War of 1812, there have been several versions of Fort Holmes (Fort George to the British) at the highest point on the Island. Some were simply tall watchtowers, with a lengthy unsafe set of stairs. These were torn down. The version I remember was built in 1936 by the Works Progress Administration (WPA) and looked pretty much like the replica of today, which was built by the State in 2015.

The differences were, the 1936 version had a couple of outbuildings in front, complete with musty dirt floor. I also remember spikes (for defense) coming out of some of the stockade logs. The palisade was accessible all the way around, not partially blocked off as it is today. The blockhouse looked pretty much the same, except you could go up to the second floor, it was not blocked off. Island kids used to grab the rope from the flagpole, and use it to swing out from the second level blockhouse windows.

~ ~ ~

One time in the early 80s I attended a City Council meeting (I didn't go to many) where the topic was the proliferation of trucks on the Island, at an alarming rate. Several Islanders, including my friend Johnny Fisher were concerned about this, and various Islanders were speaking up, and the discussion with the council became intense. Sitting to my left in a Carhartt jacket was a new fellow in town, Pat Pulte, who had just purchased the Murray Hotel. During the back and forth arguing, he nudged me and said, "See how we do things in this town?" I got a chuckle out of that. He had been here about a month, and I had been here 30 years. Ironically, there are more trucks here now, and with an extended season, than there were in the 80s.

~ ~ ~

As mentioned elsewhere, my friends and bandmates David Nephew and Don Meadows were projectionists at the Orpheum Theater, as one of their jobs for Frank Nephew. One time in summer of 1970, they found a long tube-like device with a lens on the end and a powerful lightbulb in the middle. It had no doubt been some projector paraphernalia at one time. The 1960's TV show *Batman* had been popular for several years, and was still fresh in people's minds. So Don and David made a cutout bat similar to the logo one used for the show, and attached it to the end of the lens. They would shine it at the night sky from the Orpheum, and it looked just like the Batman "bat signal". No kidding! Needless to say, people on the street passing by, were quite surprised.

Model Ships

Something I am most often recognized for, is the building of small Great Lakes freighter models. Any boy who lives around the water has probably tried his hand at homemade ship models, as well as plastic model kits. In the late 1950s-early 1960s my friend Joe Davenport and I would pencil sketch the passing freighters on paper. I also built the plastic model kits, of ocean liners and warships.

In the 1970s, I tried my hand at primitive ferry models. My materials were pine one-inch lumber and aluminium roof flashing. In the 1980s I set about improving the quality and number of the ferry boats I tried. I would gather scrap material from construction sites, and hand-paint the finished product. These were usually on $1/50^{th}$ scale. It was a "learn as you go" process. Great Lakes freighter models were hard to find, and of course I wanted to have one. So in 1991, I decided to try building my own. Once again I used scrap construction wood and flashing. The scale I found worked the best, with the 1" wood I scrounged, was $1/600^{th}$. But I would also make larger ones over time, up to $1/350^{th}$ scale.

The most popular freighter model was the Edmund Fitzgerald, followed by the Ryerson and Sykes. I sold some in a few of the Island's gift shops, and also on ebay. I also accepted custom orders. I slowly built a fan base, and sold them about as fast as I could make them. I could usually turn out four a month, when I had the spare time. I even had a Facebook page for my ships for a couple of years. One elderly gentleman from out east, Fred Dreyer, bought over 25 models from me. I even sold a freighter Joseph L. Block model to the ship's namesake, for his 90th Birthday. His granddaughter had ordered it. I branched out into Great Lakes passenger steamers, and icebreakers, and even made a couple of ocean liners, as special orders. The Chief Wawatam RR ferry became a popular model. Over the years I refined my product, for instance replacing the aluminum flashing trim with one-ply cardboard. I had found super glue did not bond wood to metal well, as a model aged.

By the 2000's my large passenger ferries had been improved too, and I was selling a few to local people, captains, and the boat lines.

One Island guy even ordered an extra large scale Huron model, with complete interiors on both decks. It was about three feet long and had to be delivered by cart. I even tried a 4-foot long model of the ferry The Straits of Mackinac, in the 80s, but never finished it. I also designed a few of my own experimental ferry models. I never lost interest in building freighters and ferries, no matter the scale. In the 21st century I again started selling them in an Island shop, a gallery on Market Street. I always kept a record of what I built, and in 33 years of Great Lakes and Straits area ship models, I have now passed 600. I did keep a few for myself.

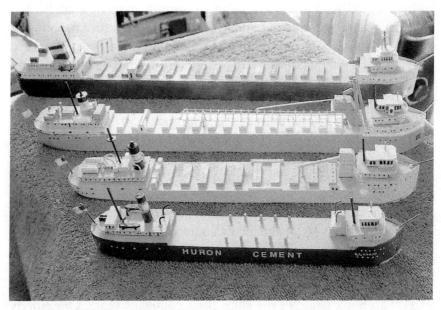

Four of Tom Chambers' more than 600 hand-built model freighters

Close Calls

I had three brushes with death during my lifetime, so far anyway. One at age 4, age 18 and age 31.

In 1957, my mom, dad and I were in Saginaw, visiting Grandma Roussin as we often did. She and her second husband Louie had a good restaurant in Saginaw, in the older area of town, in a residential district, yet close to a main thoroughfare. It was know for great homemade soup, and also homemade "friedcakes" (donuts). We had just bought a new 1956 Chevy station wagon in two-tone turquoise and white.

One day in heavy fog, we were crossing a railroad track which was not marked with the "X" crossing nor a red light. My dad stopped partially on the track, and peered both ways into the fog. Just then a familiar high-speed passenger train called "The Beeliner" was coming along, at speed. Beeliner ran all over the state, as far north as Mackinaw City, usually pulling two or three cars. It caught us on the track, shearing the engine compartment of the Chevy wagon completely off. I was in the back seat, not wearing a seat belt, and just remember hitting my head on the back of the front seat. My mom and dad sustained back injuries, which would nag them throughout life. if the car had been four feet more across the tracks, The Beeliner would have hit the passenger seats, killing all three of us instantly, I imagine.

In 1971, while the band was living the summer at Silver Birches, bass player Don Meadows and I found an old wooden kayak type boat under the lodge porch. We fished it out, and decided to test it in the lake. It fit two persons, I was in back, Don front and we had paddles. The water near shore was deep there, in the area around Pointe Aux Pins, about 8 to 10 feet. The seams in the aged wooden boat started taking on water. Abruptly, the stern went under, and being pointed, with weight in the vessel, went straight to the bottom. I had never learned how to swim (and still have not learned). Don grabbed hold of me, and pulled me in the 100 feet or so to shore. I was coughing up a bit of water. We never tried sealing up the leaks in the kayak, or used it again.

In June 1985, I was crewing aboard the new Star Line boat Nicolet, on her delivery trip from Patterson Louisiana, (near Morgan City), to St Ignace, via the Mississippi and Illinois Rivers. We had to transit many locks along the way. One such time, we needed to pull in and tie up, waiting for lock availability. I happened to be on the bow line, and we chose a high cement seawall to moor to. We had multiple 4x4 wood bumpers so the boat would incur no damage during our tie-ups.

As we neared the seawall, Tom told me, "jump!" I didn't think, I just jumped. The distance was about six feet, but I was in good physical shape that year. I landed on top of the seawall, but it was full of loose gravel and my feet slipped. My RayBan sunglasses flew off, as my feet tried to gain a footing. After a couple seconds I was ok, and secured a line. If I had slipped off the edge of the gravely seawall, I would have fallen into the river, possibly being wedged between the bow of Nicolet and the cement wall. I also could have drowned. Close call number three.

Ste. Anne's Church Square Dances

In summer 2003, my first year of working maintenance for the church, Jimmy Boynton asked me if I could fill in for Joe Scott on bass guitar for a night. Joe was an interpreter at Fort Mackinac. It would end up being a weekly summer gig for me for the next 17 years. The dances were held every Tuesday evening on the outdoor deck, from mid June to mid-August. If it was raining, we held them in the basement meeting hall. They always attracted a good crowd of summer workers, Islanders and curious tourists. Jimmy played fiddle, his mom Patti played electric piano, and John Findley, who owned Small Point B&B was the "caller." We would have guest guitarists and banjo players. Guests over the years include veterinarian Shaker Hites, Fr. Ted Brodeur on accordion, Shannon Haan Westblade, LeeAnn Ewer from the Fort, and some musicians from Beaver Island.

A typical square dance, with Jimmy Boynton calling (undated)

When John retired and moved full time to Indiana, Jimmy took over as caller. A caller basically gives instructions for the individual dance steps, and also does some singing. I only missed three square dances, between 2003-2019. In 2020, the square dances were canceled, due to the Covid-19 outbreak. Jimmy was able to bring them back, once a summer, in 2022 and 2023, for the Feast of Ste. Anne. He now had a senior post with the Jesuits in Detroit, and was no longer able to be at Ste. Anne's the whole summer. I was able to see several Island kids grow up, while dancing at the Ste Anne's square dances.

No discussion of Ste. Anne's would be complete without mentioning popular longtime priest Fr. Jim Williams. A former US Navy chaplain, Fr. Williams of Mackinaw City, came to the Island in 1992, and stayed 13 years. I had the pleasure of working for him 2003-04, as parish maintenance man. We frequently had morning political discussions over coffee at the rectory. His sermons at mass were not always popular with conservatives, but he endeared himself to many Islanders during his time on Mackinac. He ranks as one of my favorite parish clergymen, along with Fr. Terrence Donnelly , Fr. Milton Vanetvelt, and Msgr. Wilbur Gibbs.

Postscript

On November 18, 2000, I stopped consuming alcoholic beverages, after 15 years or more of indulging too much. I would not have been able to write this book, had I not done so. I would not have been able to play in a new band, had I not done so. I spent three weeks at a nice facility in Marquette. The food and the program were great, and I made some good friends. The staff even put me in charge of the PA announcements, as they determined I was the best at reading memos out loud, and the ladies in the program were happy there was at least one guy there who enjoyed badminton.

Throughout the writing of this, the question always was, what to include, and what to leave out. Also, try not to brag about anything, rather just tell the story as I remember it.

Thanks to all of you for taking time to read my recollections of life growing up on Mackinac Island. I appreciate that you made it this far.

Thomas J. Chambers

About the Author

Tom Chambers was born on Mackinac Island Michigan in 1953. He attended Mackinac Island Public School, and also spent two years at Clearwater Central Catholic High School. He attended North Central Michigan College 1970-72 and Western Michigan University 1973-75. He has spent 70 summers and 51 winters on Mackinac Island.

His first job was at the Orpheum Theater, and other jobs included the old Village Inn, Iroquois Hotel, the City of Mackinac Island, Ty's Restaurant, Lilac Tree Hotel, several painting companies and a fudge shop. He has played in two different rock bands, and even delivered ferry boats. Tom counts music, cycling, reading and photography as his hobbies. After a few years in other places, he has resided on Mackinac Island full time since April 1982.

Mackinac Island Trail Map by Chris Bessert

Just for Kids!

On a warm summer day in northern Michigan, a family of mice finds their way to Mackinac Island, the "jewel of the Great Lakes," to start a new life. Tag along and enter a world of whimsy in *A Mouse Tail on Mackinac Island*.

Join the family--Father Mouse, a lover of cheese; Mother Mouse, a singer of lullabies; Max Mouse, a collector of stamps and Millie and Maizy Mouse, sweet little babies—as they search for the perfect home. What adventures will they have? Where will they live? What dangers will they encounter? With vibrant illustrations and a charming story, *A Mouse Tail on Mackinac Island* will capture both your heart and imagination.

By Summer Porter
Illustrated by Maggie Chambers

"Soon to become a Mackinac favorite--make way for this amusing and entertaining story. Whimsical illustrations, filled with recognizable details, engage the reader throughout this thoughtful book. Who wouldn't want to join the little Mouse family on an island adventure?"
--Jennifer Powell, author and illustrator, *What I Saw on Mackinac*

"Maggie Chambers and Summer Porter have created a fun and colorful adventure of a family of mice all readers will enjoy--and, like a bite of island fudge, *A Mouse Tail on Mackinac Island* will prove very satisfying!"
--Jim Bolone, co-author, *The Dockporter*

"This charming children's book captures the essence of Mackinac Island. A classic 'tail' for people of all ages that will spark a sense of wonder for years to come."
--Mayor Margaret Doud, city of Mackinac Island

Learn more at www.MouseTailOnMackinac.com

Printed in the USA
CPSIA information can be obtained
at www.ICGtesting.com
LVHW010006061024
792988LV00004B/81